Methods in Environmental Geology

Series editor: Federal Institute for Geosciences and Natural Resources (BGR)
Stilleweg 2, 30655 Hannover, Germany

This book was written in the course of the project "Methods for Studying the Subsurface of Planned, Operating, and Abandoned Waste Disposal Sites". It was financed by the German Federal Ministry for Education, Science and Technology (BMBF); Project coordination: Federal Environmental Agency (UBA) / Research Grant-No. 1460605; Project leadership: Federal Institute for Geosciences and Natural Resources (BGR), Germany. The English version of the book was edited with the help of the U.S. Geological Survey (USGS).

The authors are responsible for the content of their contributions.

Springer

Berlin
Heidelberg
New York
Barcelona
Hong Kong
London
Milan
Paris
Singapore
Tokyo

Friedrich Kuehn · Trude V.V. King · Bernhard Hoerig · Douglas C. Peters (Eds.)

Remote Sensing for Site Characterization

With 11 Tables and 117 Figures, 70 in colour

Springer

Editors

Friedrich Kuehn
Federal Institute for Geosciences and
Natural Resources (BGR), Berlin Office
Wilhelmstraße 25–30
13593 Berlin, Germany

Bernhard Hoerig
Federal Institute for Geosciences and
Natural Resources (BGR), Berlin Office
Wilhelmstraße 25–30
13593 Berlin, Germany

Trude V. V. King
U.S. Geological Survey (USGS)
Denver Federal Center
PO Box 25046, MS 964
Denver CO 80225-0046, USA

Douglas C. Peters
Peters Geosciences
169 Quaker St.
Golden, CO 80401-5543, USA

Translators

Clark Newcomb
Federal Institute for Geosciences and
Natural Resources (BGR)
Stilleweg 2
30655 Hannover, Germany

Henry Toms
Federal Institute for Geosciences and
Natural Resources (BGR)
Stilleweg 2
30655 Hannover, Germany

The scientific contributions of the U.S. Geological Survey employees are not subject to copyright.

ISSN 1438-8863
ISBN 3-540-63469-X Springer-Verlag Berlin Heidelberg New York

Library of Congress Cataloging-in-Publication Data
Remote sensing for site characterization / Friedrich Kuehn ... [et al.].
p. cm. -- (Methods in environmental geology, ISSN 1438-8863)
ISBN 3-540-63469-X (Hardcover : alk. paper)
1. Environmental geology--Remote sensing. 2. Environmental geology--Remote sensing--Case studies. I. Kuehn, Friedrich, 1948- II. Series.

QE38 .R46 2000
550'.28--dc21   99-053337

This work is subject to copyright. All rights are reserved, whether the whole or part of the material is concerned, specifically the rights of translation, reprinting, reuse of illustrations, recitation, broadcasting, reproduction on microfilms or in any other way, and storage in data banks. Duplication of this publication or parts thereof is permitted only under the provisions of the German Copyright Law of September 9, 1965, in its current version, and permission for use must always be obtained from Springer-Verlag. Violations are liable for prosecution under the German Copyright Law.

© Springer-Verlag Berlin Heidelberg 2000
Printed in Germany

The use of general descriptive names, registered names, trademarks, etc. in this publication does not imply, even in the absence of a specific statement, that such names are exempt from the relevant protective laws and regulations and therefore free for general use.

Cover Design: Erich Kirchner, Heidelberg
Dataconversion: Büro Stasch, Bayreuth

SPIN: 10531532   30/3136 – 5 4 3 2 1 0 – Printed on acid-free paper

Contents

1 Introduction 1

2 Remote Sensing: An Overview of Physical Fundamentals 4

3 Obtaining Remote-sensing Data 9
3.1 Satellite-based Methods 9
3.2 Aircraft-based Methods 11
    3.2.1 Aerial Photographs 11
    3.2.2 Nonphotographic Imaging Sensors 20

4 The Use of Remote Sensing in Waste Disposal Site Investigation .. 33
4.1 Investigative Objectives and Interpretative Criteria 33
4.2 Case Studies 38
    4.2.1 Characterization of a Waste Disposal Site 38
    4.2.2 Exploring the Immediate Vicinity Around a Waste Disposal Site .. 44
    4.2.3 Subsurface Characteristics of a Waste Disposal Site 55
    4.2.4 The Search for New Waste Disposal Sites 58

5 Verification of Remotely Sensed Data 59
5.1 Introduction 59
5.2 Virtual Versus In-Situ Verification 59
5.3 Verification of Vegetation Data 62

6 Case Studies 63
6.1 Introduction 63
6.2 Archival Aerial Photographs Used to Evaluate the Subsurface
    of Waste Disposal Sites (Arnstadt, Germany) 64
    6.2.1 Introduction and Problem Description 64
    6.2.2 Geophysical Investigations 66
    6.2.3 Interpretation of Aerial Photographs 68
    6.2.4 Summary 73
6.3 Airborne Remote Sensing to Characterize Waste Disposal Sites
    (Schoeneiche, Germany) 73
    6.3.1 Introduction and Problem Description 73
    6.3.2 Interpretation of Aerial Photographs and Scanner Images .. 76
    6.3.3 Summary 93

6.4	Thermal Remote Sensing to Detect Buried Waste Material (Oak Ridge, U.S.A.)	96
	6.4.1 Introduction	96
	6.4.2 Background	96
	6.4.3 Imagery Analysis	99
	6.4.4 Ground Data	103
	6.4.5 Conclusions	104
6.5	Remote Sensing for Monitoring the Effects of Mining in Sudbury, Canada	106
	6.5.1 Introduction	106
	6.5.2 Sudbury Case Study	106
	6.5.3 Summary	113
6.6	Multispectral Remote Sensing to Characterize Mine Waste (Cripple Creek and Goldfield, U.S.A.)	113
	6.6.1 Introduction	113
	6.6.2 Investigation Methodology	116
	6.6.3 Case Studies	117
	6.6.4 Prioritizing Waste Site Investigations Based on Remote Sensing	161
	6.6.5 Summary	163
6.7	Applications of Imaging Spectroscopy Data: A Case Study at Summitville, Colorado	164
	6.7.1 Introduction	164
	6.7.2 Imaging Spectrometer Data	165
	6.7.3 Data Analysis	167
	6.7.4 Verification of Imaging Spectrometer Data and Results	168
	6.7.5 Mapping Minerals	169
	6.7.6 Mapping Vegetation	175
	6.7.7 Senescence/Stress Mapping	182
	6.7.8 Conclusions	184

Epilogue ... 187

Acknowledgements ... 189

Sources of Maps, Photos, and Images ... 191

Glossary of Frequently Used Abbreviations ... 193

References ... 195

Additional References ... 203

Subject Index ... 207

List of Contributors

Clark, Roger N.
U.S. Geological Survey (USGS), Denver Federal Center
PO Box 25046, MS 964
Denver CO 80225-0046
USA

Evers, Thomas K.
Oak Ridge National Laboratory
PO Box 2008
Oak Ridge, TN 37831-6274
USA

Hauff, Phoebe L.
Spectral International Inc.
PO Box 1027
Arvada, CO 80001
USA

Hoerig, Bernhard
Federal Institute for Geosciences and Natural Resources (BGR)
Berlin Office
Wilhelmstraße 25–30
13593 Berlin
Germany

Huff, Dale
Oak Ridge National Laboratory
PO Box 2008
Oak Ridge, TN 37831-6274
USA

Irvine, John M.
Science Applications International Corporation (SAIC)
4001 North Fairfax Drive, Suite 725
Arlington, VA 22203
USA

King, Amy L.
Oak Ridge National Laboratory
PO Box 2008
Oak Ridge, TN 37831-6274
USA

King, Trude V. V.
U.S. Geological Survey (USGS)
Denver Federal Center
PO Box 25046, MS 964
Denver CO 80225-0046
USA

Kuehn, Friedrich
Federal Institute for Geosciences and Natural Resources (BGR)
Berlin Office
Wilhelmstraße 25–30
13593 Berlin
Germany

Odenweller, Julie
Environmental Research Institute of Michigan (ERIM)
PO Box 134001
Ann Arbor, MI 48113-4001
USA

Peters, Douglas C.
Peters Geosciences
169 Quaker St.
Golden, CO 80401-5543
USA

Schmidt, Dietmar
Deutsche PhoneSat AG
FEZ Potsdam GmbH
Berliner Straße 50
14464 Berlin
Germany

Singhroy, Vernon
Canada Centre for Remote Sensing (CCRS)
588 Booth Street, Room 207
Ottawa, Ontario
Canada K1A 0E4

Smyre, John L.
Oak Ridge National Laboratory
PO Box 2008
Oak Ridge, TN 37831-6274
USA

Stahl, Gary
Environmental Research Institute of Michigan (ERIM)
PO Box 134001
Ann Arbor, MI 48113-4001
USA

Swayze, Gregg A.
U.S. Geological Survey (USGS)
Denver Federal Center
PO Box 25046, MS 964
Denver CO 80225-0046
USA

Chapter 1

Introduction

Friedrich Kuehn

The Federal Institute for Geosciences and Natural Resources (BGR), Germany, in cooperation with scientists from universities, research institutes, and industry, has characterized waste disposal sites in Germany. The work was commissioned by the Federal Environmental Agency (Umweltbundesamt) and funded by the German Federal Ministry for Education, Science, Research and Technology (BMBF) as a joint research study called "Methods for the Investigation and Characterization of the Ground below Waste Disposal Sites".

The primary objective of the study was to increase the understanding of the subsurface characteristics of waste disposal sites using geologic and geophysical methods. The work focused on determination of the thickness and physical/chemical properties of the lining and underlying rocks, and identification of potential migration paths for fluids in the vicinity of disposal sites.

This handbook, "Remote Sensing for Site Characterization", is the first in a series of planned publications under the general title "Methods in Environmental Geology". It is based on the German edition (Kuehn and Hoerig 1995), which has been considerably expanded by the addition of sections by U.S. and Canadian scientists. This volume illustrates the use of theoretical and technical aspects of airborne and satellite-based methods in the study of waste sites and gives examples of their applications.

In recent years, remote-sensing methods have been increasingly recognized as a means of obtaining crucial geoscientific data for both regional and site-specific investigations. Remote-sensing data provides a synoptic perspective not achievable with traditional field studies. These methods are effective for basic and applied research covering a wide range of subjects, including mineral exploration and geo-environmental evaluation.

Remote-sensing methods can provide geoscientific data for large areas in a relatively short time. Remote sensing is an excellent site characterization tool because it is not limited by extremes in terrain or hazardous conditions, which may be encountered during an on-site appraisal. Whenever possible, remote-sensing data should be acquired and integrated into the early stages of an investigation and used in conjunction with traditional mapping techniques. Such data is best suited for the following purposes:

- Preliminary assessment and site characterization of an area prior to the application of more costly and time consuming traditional assessment techniques, such as field mapping, drilling, and geophysical surveys.

- Clarification of geoscientific problems using the broad perspective provided by an aircraft or satellite image.
- Geoscientific assessment of regions with limited or no access, such as rugged terrain, hazardous sites, and disaster areas.

The data obtained from satellite-based remote-sensing systems is best suited for regional studies at scales of 1 : 500 000 to 1 : 100 000 and, in some cases, 1 : 25 000. These data are commonly used to characterize natural resources that have a wide distribution (e.g., tropical rain forests), to monitor widespread altering of the landscape (e.g., desertification, coastal changes, ice cover of polar waters), as well as detect and monitor environmental problems (e.g., forest fires and oil spills).

Satellite images have also been shown to be an effective tool for characterizing and assessing areas of human activity, such as open-pit mines and military training areas. Singhroy (Section 6.5, this volume) and Peters and Hauff (Section 6.6, this volume) provide examples of the successful use of satellite data to detect changes in landscape character and general mineralogy. However, for detailed site characterization, satellite data often is of limited use because of its relatively low spatial resolution.

For detailed geoenvironmental assessment of small-scale hazardous waste sites, waste dumps, and other areas altered by humans, data with greater spatial resolution is needed. For most geoenvironmental problems, mapping scales of 1 : 10 000 or larger are required to sufficiently characterize sites of limited size. Both high-resolution aerial photographs and airborne scanners provide data at the required spatial resolution (e.g., 50 cm and better).

To obtain maximum scientific return and eliminate questions regarding the utility of remote-sensing data, the inexperienced user will benefit from working with a remote-sensing expert. Involvement of remote-sensing experts insures that the most up-to-date methods and techniques are applied to the data. It takes the experience of a remote-sensing expert to recognize the full potential of the aerial and satellite image data. Prior to beginning a remote-sensing project, it is beneficial to first define the objectives and goals of the study so that the most suitable techniques and methods can be used. In some cases, this analysis may determine that remote-sensing methods are inappropriate for the investigation.

This handbook is designed to provide examples of the application and limitations of remote-sensing data to investigate waste disposal and mining sites. It is not intended to be used as a textbook, but, instead, to provide insights into the application and limits of remote-sensing data by discussion of case studies. The selection of case studies presented here should not be viewed as the only applications to mining, waste disposal, or other problems of environmental concern. They should serve as examples of the methodology for extrapolation to other geoenvironmental concerns.

In Chapter 6, we provide case studies to demonstrate the possible uses of remote-sensing data to assess waste disposal and mining sites. Case studies from the United States (Summitville and Cripple Creek, Colorado; Goldfield, Nevada; and Oak Ridge, Tennessee), Canada (Sudbury, Ontario), and Germany (Schoeneiche, Brandenburg; and Arnstadt, Thuringia) are included.

Comparison of case studies from Germany and North America depict how geographical differences alone can influence the selection and application of site characterization methods. Differences in population density and distance between sites of interest are shown to have a significant impact on the selection of methods used.

The authors and editors of this volume believe that the case studies in this volume will illustrate the advantages of applying remote-sensing techniques to decision makers faced with their own environmental investigations. The consistent and knowledgeable application of remote-sensing methods will improve the timeliness, cost-effectiveness, and thoroughness of most environmental site assessments.

Chapter 2

Remote Sensing:
An Overview of Physical Fundamentals

Bernhard Hoerig · Friedrich Kuehn

Remote sensing refers to specific methods used for obtaining information about the Earth's surface. These methods, which sense electromagnetic (EM) radiation, have no direct contact between the sensor – carried by either an aircraft or satellite – and the object(s) being observed. Remote sensing utilizes EM radiation principally in the ultraviolet, visible light, infrared, and microwave portions of the EM spectrum. Single or multichannel data acquisition systems are used as tools for gathering remotely sensed data. Airborne geophysical methods, such as airborne electromagnetics, are not considered as remote-sensing methods, although they meet the above description.

Remote-sensing techniques are divided into active and passive methods. Passive methods use reflected solar radiation and radiation emitted from a surface. Active remote-sensing systems provide a source of radiation (e.g., radar). Active systems will not be discussed in detail in this book due to their limited use for investigating waste disposal sites (Section 3.2.2.5), although high-resolution digital elevation models (DEMs) generated from airborne radar interferometry and laser scanning may find applications in the future.

Interaction of electromagnetic radiation with the Earth's surface provides information about the reflecting or absorbing surface. Due to the high frequencies of the EM radiation (micrometer to nannometer range), there is limited penetration of the targeted objects. Consequently, data and images are obtained only from the surface. Remote-sensing information on conditions and structures underlying a natural or artificial terrain surface can be derived only by inference. Thus, the reliability of an interpretation depends on the knowledge and experience of the interpreter. An interpreter of aerial photographs and other remote-sensing data should have a thorough understanding of the characteristics of landscapes, possible subsurface conditions, and how these factors interact and affect the resulting data.

A map derived from the interpretation of aerial photographs and other remote-sensing data is influenced by subjective factors. Maps generated on the basis of automatic classification techniques depend on the quality and appropriateness of the input data and analysis techniques used. Therefore, it is particularly important to check interpretations of remote-sensing data in the field. Field or ground checks may be necessary at the start and during a remote-sensing project to establish a key for interpreting the data or to check intermediate interpretations (see Chapter 5 and 6 of this volume).

In the following sections, there is repeated reference to EM radiation as a carrier of remote-sensing information. Therefore, it is important to define the differ-

Table 2.1. Spectral range of electromagnetic radiation utilized by remote-sensing sensors; spectral divisions by Erb (1989)

Radiation	Abbreviation	Wavelength λ (µm)
Near ultraviolet	NUV	0.315 – 0.38
Visible light	VIS	0.38 – 0.78
Near infrared	NIR-I	0.78 – 1.4
	NIR-II	1.4 – 3.0
Middle infrared	MIR	3.0 – 50.0
(thermal infrared)	MIR-I	3.0 – 5.5
	MIR-II	8.0 – 15.0
Far infrared	FIR	50.0 – 1 000
Microwave (radar)	MW	1 000 – 1×10^6

ent ranges of the EM spectrum. In the literature, it is difficult to find an unambiguous division of the electromagnetic spectrum, because the EM spectrum is continuous and does not have natural divisions. Divisions are commonly based on type of sensor, response of natural materials, etc. We decided to use the divisions of the spectrum by Erb (1989) as for discussing the sensors and systems. In addition to Erb's divisions, we will use a further subdivision of the **middle infrared** (MIR) into MIR-I and MIR-II (see Table 2.1) for remote-sensing purposes.

The sun is the main source of the Earth's incident EM radiation. The energy spectrum of the sun is approximately identical to that of a blackbody at 5 900 K. Solar radiation reflected from the Earth's surface is in the NUV to NIR-II (0.315–3.0 µm). However, there are gaps in this spectral range due to scattering, absorption, and reflection by gases, particulates, and other atmospheric constituents (Fig. 2.1). Comparison of the energy spectrum before and after passage through the atmosphere shows that gaps result from absorption by atmospheric water and carbon dioxide at 1.4 µm, 1.9 µm, and 2.5–3.0 µm. Little or no solar radiation reaches the Earth's surface at these wavelengths and generally cannot be used for remote-sensing purposes.

Solar radiation can be absorbed or reflected by a surface, depending on the composition and properties of the materials. The reflections can be either direct or diffuse. The proportions of reflection, absorption, and penetration of the incident radiation at the Earth's surface depend on the physical, structural, and textural properties of the surface. Molecules and chemical bonds (which make up all materials, including soils, rocks, water, and plants) are characterized by a specific energy state. The absorption of incident radiation provides the energy required for the transition from one energy state to another. Remote-sensing systems record the radiation after its interaction with the Earth's surface; hence, the properties of the soil, rock, and other materials at the Earth's surface are indirectly recorded in the data or images.

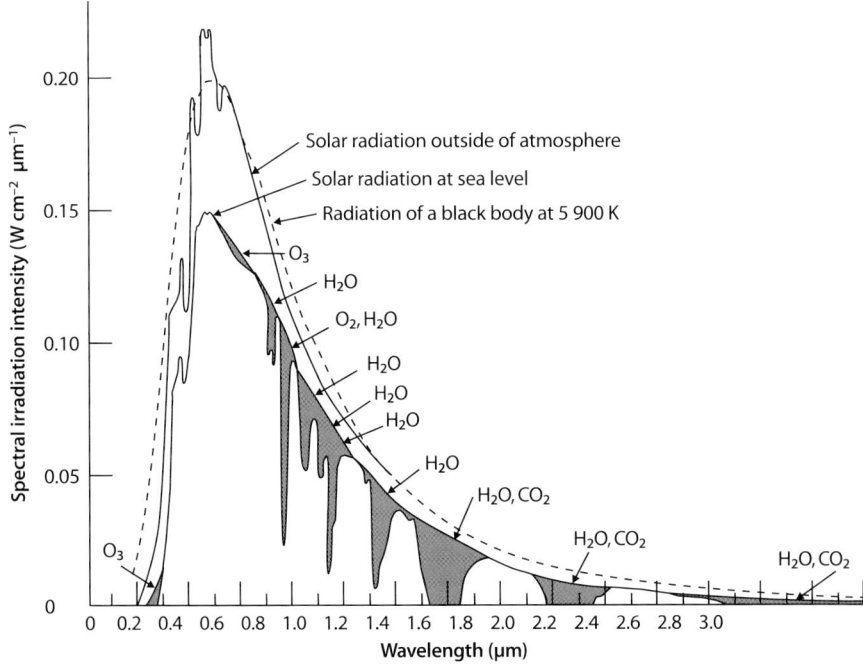

Fig. 2.1. Energy spectra of incoming solar radiation before and after passing through the atmosphere, and the radiation emitted by a blackbody at 5 900 K; the gaps in the spectrum caused by absorption in the atmosphere are shown (after Kronberg 1985)

In practical application, thermal radiation is in the approximate spectral range of 3 to 50 µm (theoretically up to 1 000 µm). In this range of wavelengths, the surface of the Earth approximates a radiating blackbody. A blackbody absorbs all incident radiation. However, the spectrum of a blackbody is a function of its temperature, as given by the *Stefan-Boltzmann law* (see Gupta 1991):

$$W = \sigma T^4 \,, \tag{2.1}$$

where
- W = spectral radiation emission in $W\,m^{-2}$
- σ = Stefan-Boltzmann constant
- T = temperature of the object.

As Fig. 2.2 shows, the maximum radiation of a blackbody shifts towards shorter wavelengths with increasing temperature. The temperature-dependent radiation peak of a blackbody provides an opportunity for the application of EM radiation in remote sensing. Because of its "apparent" temperature of about 6 000 K, the maximum energy of solar radiation is within the visible part of the spectrum (VIS) at 0.48 µm. The maximum radiation from the much colder surface of the Earth,

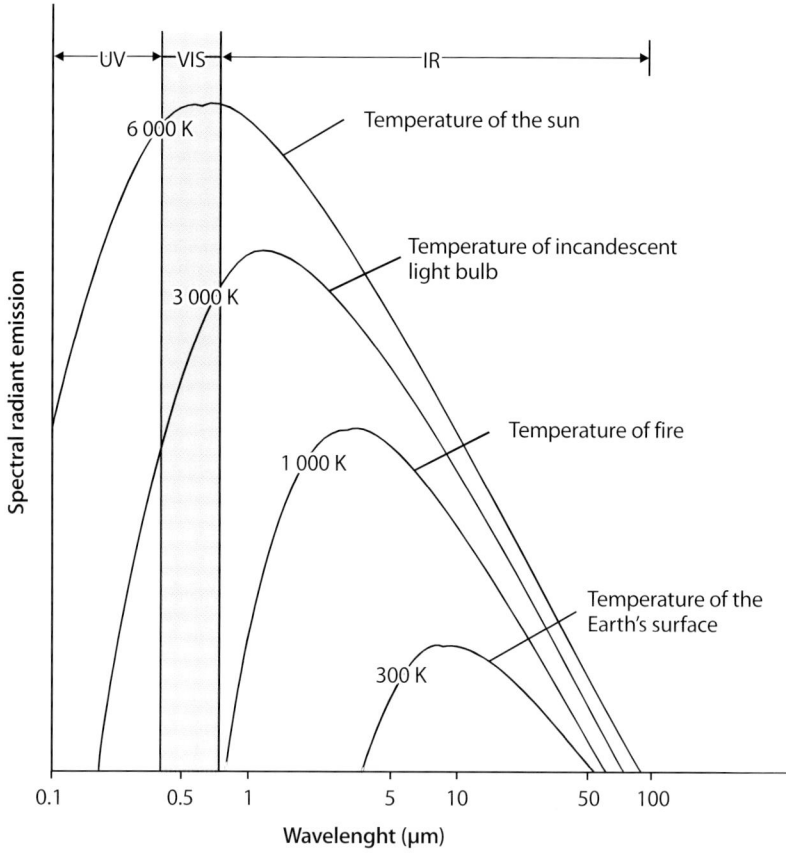

Fig. 2.2. Energy spectra of the radiation emitted by the sun, a light bulb, a fire, and the Earth's surface, showing the shift to higher wavelengths with lower temperatures (after Gupta 1991)

with an average temperature of 300 K, is at about 9.7 µm. Thus, the best range for recording thermal radiation from the Earth's surface is between 8 and 12 µm.

Within this spectral range, slight temperature variations resulting from differences in the type of soil, soil moisture, or the presence of pollutants can commonly be identified. Therefore, thermal scanners are designed to be the most sensitive in the spectral range of 8–12 µm. To investigate hot surfaces (e.g., radiation from a blast furnace or the structure of a lava flow), a spectral range of 3–5 µm is more suitable.

Radiation in the visible and infrared portions of the spectrum is absorbed by certain constituents of the atmosphere. Major absorption bands lie between 5 and 8 µm and between 13 and 30 µm. Accordingly, these ranges of the electromagnetic spectrum are not generally used for remote sensing. Applications of thermal radiation are described in Chapter 4 and case studies are given in Sections 6.3 and 6.4.

Chapter 3
Obtaining Remote-sensing Data

Friedrich Kuehn · Bernhard Hoerig

3.1
Satellite-based Methods

Images from satellite-based remote-sensing systems are primarily used to obtain data on large areas (Barrett and Curtis 1992; Theilen-Willige 1993). The spatial resolution of a satellite-based remote-sensing system is commonly insufficient to delineate and identify small objects on the Earth surface (see Table 3.1). Satellite images (*Landsat Thematic Mapper, SPOT, MOMS, ERS-1, ERS-2, RADARSAT*, etc.) can provide data on the extent and structure of waste disposal sites and characterize the landscape features of a given site (Fig. 3.1). However, the highest resolution and most informative data for waste disposal site investigation are obtained from airborne remote-sensing systems.

Satellite-based imaging systems operate much like aircraft-based systems with the exception that there is a greater distance between the sensor and the object being evaluated and the methods of data transfer are different. However, in general, satellite-based systems are analogous to the aircraft-based systems described in Section 3.2.

Satellite photographs and aerial photography can be interpreted using the same methods of analysis. Satellite photographs differ from aerial photographs in that the former have lower spatial resolution and larger field of view. Overlapping satellite photographs can be stereoscopically evaluated in a manner analogous to aerial photographs. The images are available from specialized marketing companies as copies on paper or film (Strathmann 1993). For planning purposes, digitized and rectified photographs can be obtained on request at the scale of standard topographic maps up to a scale of 1 : 10 000.

Data from nonphotographic systems (*Landsat Thematic Mapper, SPOT, ERS-1, ERS-2, RADARSAT* etc.) may be obtained in digital form on magnetic tape, optical disk, or CD-ROM or as hard copy prepared after standardized image processing. Experience has shown that the optimum results are achieved when the end user contributes to the planning and is involved in the data analysis. However, access to specialized software is necessary for this level of involvement.

There are some limitations on the use of satellite data. The data commonly have to be obtained from an archive and these data often do not represent the current conditions at a given site. For geologic applications, data acquisition is sometimes restricted to a particular season and the study area must be cloud free. Thus, the chances of success are somewhat reduced, because the *Landsat Thematic Mapper*, for example, only covers the same site on the Earth every 16 days and may not co-

Table 3.1. Overview of the major commercial satellite remote-sensing systems

Satellite (Country)	Spatial resolution (pixel size)	System	Spectral range
Landsat Thematic Mapper (TM) (USA)	30 m × 30 m	multispectral scanner[b]	0.45 – 0.52 µm 0.52 – 0.60 µm 0.63 – 0.69 µm 0.76 – 0.90 µm 1.55 – 1.75 µm 2.08 – 2.35 µm
	120 m TIR[a]		10.40 – 12.50 µm
Spot (France)	20 m × 20 m (multispectral)	multispectral scanner	0.50 – 0.59 µm 0.61 – 0.68 µm 0.79 – 0.89 µm
	10 m × 10 m (panchromatic)	scanner	0.51 – 0.73 µm
KVR-3000 (Russia)	1.5 m × 1.5 m	photographic camera	panchromatic
Seasat (USA)	25 m × 25 m	radar[b]	23.5 cm
ERS-1 and 2 (EU)	12.5 m × 12.5 m	radar[b]	3.6 cm
Radarsat (Canada)	8.3 m × 8.4 m	radar[b]	5.6 cm (fine mode)
JERS-1[c] (Japan)	18 m × 18 m	radar[b]	23.0 cm
	18 m × 24 m	multispectral scanner[b]	0.52 – 0.60 µm 0.63 – 0.69 µm 0.76 – 0.86 µm[d] 0.76 – 0.86 µm[d] 1.60 – 1.71 µm 2.01 – 2.12 µm 2.13 – 2.25 µm 2.27 – 2.40 µm

[a] TIR: thermal infrared;
[b] For technical details, see Section 3.2.2;
[c] After Earth Resources Satellite Data Analysis Center (ERSDAC);
[d] Two images for a stereopair.

incide with optimum weather conditions. Bad weather might result in delaying data acquisition for a year or require the use of archival material, which commonly happens in Central Europe.

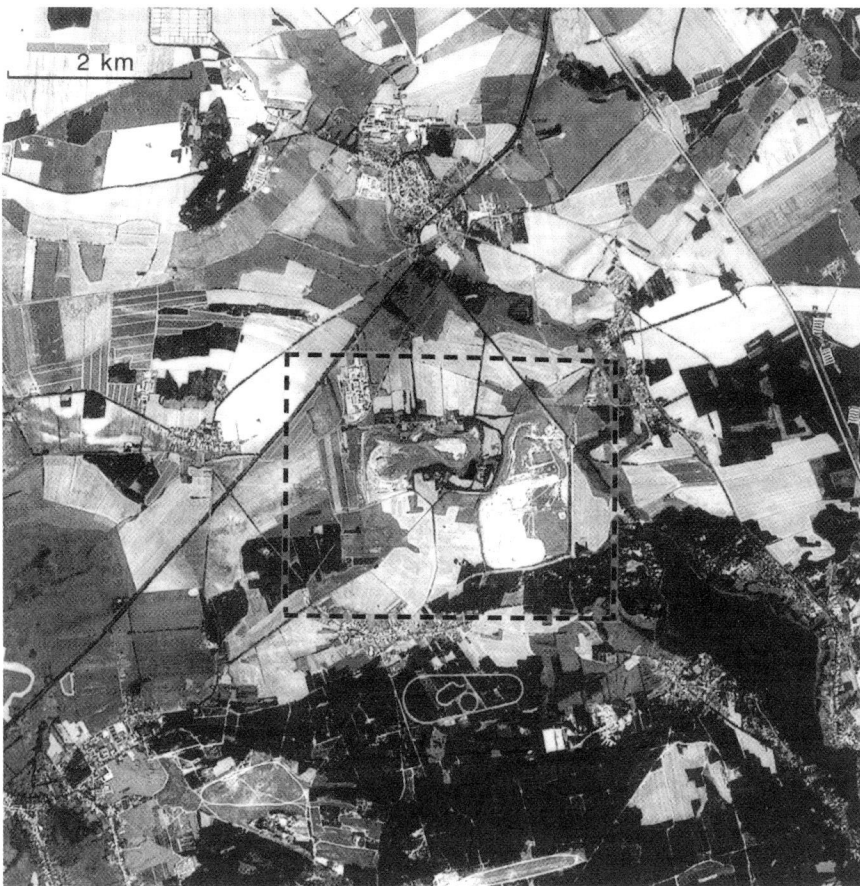

Fig. 3.1. Schoeneiche and Schoeneicher Plan waste disposal sites (*inside dashed line*), south of Mittenwalde near Berlin, on a *SPOT* image taken on August 25, 1990; panchromatic module (see Table 3.1) courtesy of FPK Ingenieurbüro für Fernerkundung, Photogrammetrie und Kartographie GbR (Consultants for Remote Sensing, Photogrammetry and Cartography) in Berlin

3.2
Aircraft-based Methods

3.2.1
Aerial Photographs

Despite all technical progress in digital imaging, standard aerial photographic film remains an important remote-sensing tool. The cost of aerial photography is relatively low, the data are informative and easy to manage, and the film does not require special image processing systems for analysis. Aerial photography can address a multitude of geoscientific questions and can be highly effective when used for logistics and planning (Boeker and Kuehn 1992).

Cameras for aerial photography, also referred to as metric cameras and frame reconnaissance cameras, are analogous to standard photographic cameras in that they use lenses, shutters, and film. However, these cameras are more expensive, as well as mechanically and electronically complicated, because of the need to adjust the orientation of the camera during data acquisition. Ideally, aerial photography equipment is coupled to a Global Positioning System (GPS) to ensure more precision.

Photographic multispectral cameras were used in planes and satellites in the 1970s and 1980s (Colwell 1983; Kuehn and Oleikiewitz 1983). These cameras are now no longer in use for several reasons: their complicated mode of operation, spectral distortion depending on the angle of incidence on the camera lenses and filters, and the complicated and imprecise evaluation procedures. They have been replaced by multispectral digital imaging systems.

Vertical aerial photographs provide a three-dimensional impression of an imaged area. The evaluation of stereopair photographs requires an overlap of 60–90% with the successive image (see Figs. 3.2 and 3.3). An overlap of approximately 35% is needed between adjacent flight lines to allow for flight irregularities. As the

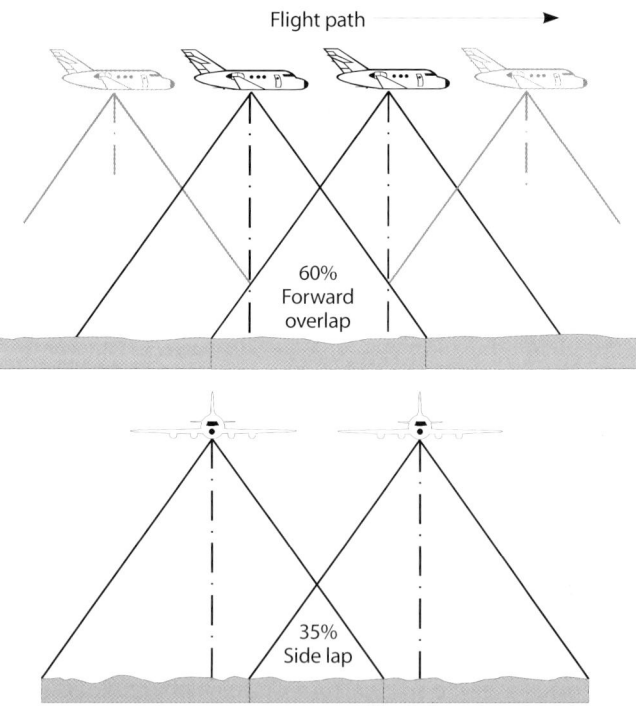

Fig. 3.2. Flight with a 60% forward overlap and a 35% flight-path overlap (side lap) are necessary for effective stereoscopic evaluation (3-D) of photograph pairs (after Schneider 1974)

distance between any two consecutive images is considerably greater than the distance between the human eyes, the relief of the target area appears strongly enhanced when the images are viewed stereoscopically. This enhancement commonly provides detailed information for geological and environmental analysis. The aerial photographs can be used for, but are not limited to detecting man-made features, for example, hollows filled with waste, and gentle depressions where ground water and leachate from waste disposal sites can accumulate, for characterizing changes in relief indicative of the presence of fractures, and for delineating surface depressions resulting from subsurface mining or subrosion.

In addition, aerial photographs can often be analyzed to identify rock formations and soil characteristics via changes in color or gray scale, typical relief forms, distinctive vegetation types, characteristic drainage patterns and specific types of land use. In general, as the incidence of man-made features increases in an area, it becomes more difficult to extract geologic information from aerial photographs. However, in the cases of environmental assessment it is difficult to avoid working in areas affected by man-made features because soil and ground-water pollution are often direct consequences of population growth and industrial expansion.

Vertical aerial photographs are commonly produced at scales of 1 : 20 000 to 1 : 5 000 for the purposes of environmental geology. Using a lens with a focal length of 150 mm, the flight altitude required to produce these scales is between 3 000 m and 750 m, respectively, above the surface. Standard film size of 23 cm × 23 cm then will capture an area of 4 600 m × 4 600 m (at a scale 1 : 20 000) or 1 150 m × 1 150 m (scale 1 : 5 000). The basic parameters of aerial photography, i.e., the scale (S) of the aerial photograph, the field of view (A) of the camera, and the ground resolution (R_g), can be calculated using the following equations (Bormann 1981a):

$$S = \frac{c_f}{h_g} = \frac{1}{S_n} \tag{3.1}$$

$$A = \frac{h_g k}{c_f} \tag{3.2}$$

$$R_g = \frac{h_g k}{R_L} \tag{3.3}$$

where
- c_f = is the focal length of the camera lens,
- S_n = scale constant,
- h_g = flight altitude over terrain,
- k = film frame size (usually 23 cm × 23 cm), and
- R_L = line resolution (lpi) of the film according to the manufacturer.

Equation 3.3 gives an estimate of the spatial resolution, R_g. To precisely calculate the resolution of a given "film–lens" system, a contrast modulation function has to be used which describes the relation between the degree of contrast in the terrain and in the image as a function of object size for a given spatial frequency domain (Bormann 1981b).

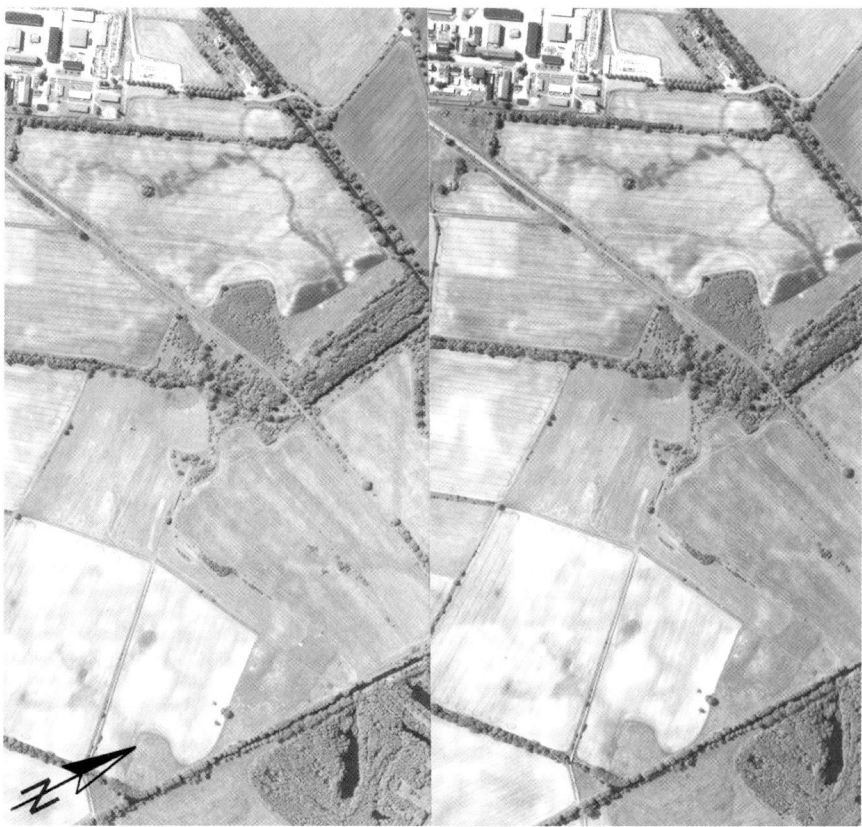

Fig. 3.3. Stereopair of black-and-white photographs showing an abandoned waste disposal site now used for agriculture (*center right*; visible with pocket stereoscope). The meanders of a former stream (*near top of photo*) and nearby filled bomb crater (vegetation) can be seen (source: Landesvermessungsamt Brandenburg)

Panchromatic black-and-white film. Depending on the task and the desired results, different kinds of film can be used for aerial photography. The most commonly used film for aerial photographs is black-and-white panchromatic film (Schneider 1974). This film is frequently used for geodetic and cartographic purposes, but may also be used for specific thematic problems. The high spatial resolution allows minor joints or fractures to be mapped. Phenomena such as changes in rock and soil characteristics, soil moisture, or secondary effects related to contamination can also be detected with this film. Stereoscopic techniques using pairs of aerial photographs allow an experienced interpreter to characterize objects at the surface of a site in detail (see Fig. 3.3). Schneider (1974), Kronberg (1984) and Ciciarelli (1991) show a multitude of applications of panchromatic photographic film.

Infrared film. Black-and-white film can be sensitized to extend beyond the visible range (VIS) into near-infrared wavelengths (NIR-I). This part of the near-infrared spectrum, also referred to as the photographic infrared, is the range of wavelengths from 0.78 μm to about 0.9 μm (Table 2.1). The EM radiation detected by near-infrared film is strictly reflected solar radiation and should not be confused with the thermal infrared region. The latter is the EM range between 8 and 12 μm and is used for thermal imaging (Table 2.1). Near-infrared film can be used to detect changes in soil moisture and the extent and state of plant cover. Due to the intense absorption of infrared wavelengths by water, near-infrared film is highly sensitive to changes in soil moisture content. Conversely, healthy vegetation intensely reflects radiation in the same range. Color infrared film has largely replaced black-and-white infrared film for mapping biotopes and evaluating the vitality or health of vegetation.

Color film and color-infrared film. Color aerial photographs can be used to depict a terrain in natural colors (see Fig. 3.5). Color film consists of three light-sensitive layers for blue, green, and red. By substituting a near-infrared-sensitive layer for the blue-sensitive layer, a false color photograph can be produced. The near-infrared-sensitive layer reacts to the intense reflection of NIR-I radiation by vegetation. As a result, if a reversal process is used for developing the film, the chlorophyll-rich vegetation will be represented by intense red colors and a low chlorophyll content by pale greyish red. Color-infrared film also results in sharper images than normal color film with greater contrast from high altitudes due to the lower dispersion of infrared radiation in the atmosphere. The increase in contrast is enhanced by the use of a yellow filter to suppress the blue frequencies.

Color-infrared (CIR) film has proved indispensable for a broad range of thematic mapping applications. Environmental mapping relies on CIR aerial photography because of its high sensitivity to plant vitality. The methods for mapping the vitality of trees in the vicinity of a waste disposal site using CIR images are described in Chapter 6 of this volume (Section 6.3.2.3).

A CIR aerial photograph of the Vorketzin waste disposal site and adjacent areas west of Berlin taken on July 28, 1990, is shown in Fig. 3.4. Prior to WWII, the abandoned, water-filled clay pits of the brickyards in the area were used as disposal sites. A geoenvironmental assessment of the region was initiated to determine potential migration paths of pollutants toward an unconfined aquifer subsequent to the removal of a large volume of clay.

CIR aerial photographs of the Vorketzin site provide a wide array of information about the waste disposal site. Possible pollutant pathways from the disposal site into the surrounding areas can be traced by mapping damaged vegetation on the basis of changes in color (from deep red to light gray). Changes in the color of water in the adjacent ponds (former clay pits) are explained by different degrees of pollution. The blue in the left portion of the photograph depicts water with elevated turbidity (groundchecked). The discoloration was interpreted as due to suspended matter transported in the ground water from buried waste in the landfill. Ponds isolated from the nutrient-rich ponds by impermeable clay barriers are visible in lower left corner of the photograph. The black color of these water bodies, like those on the northern edge of the photo, indicate low contents of suspended matter (cf. Fig. 4.14).

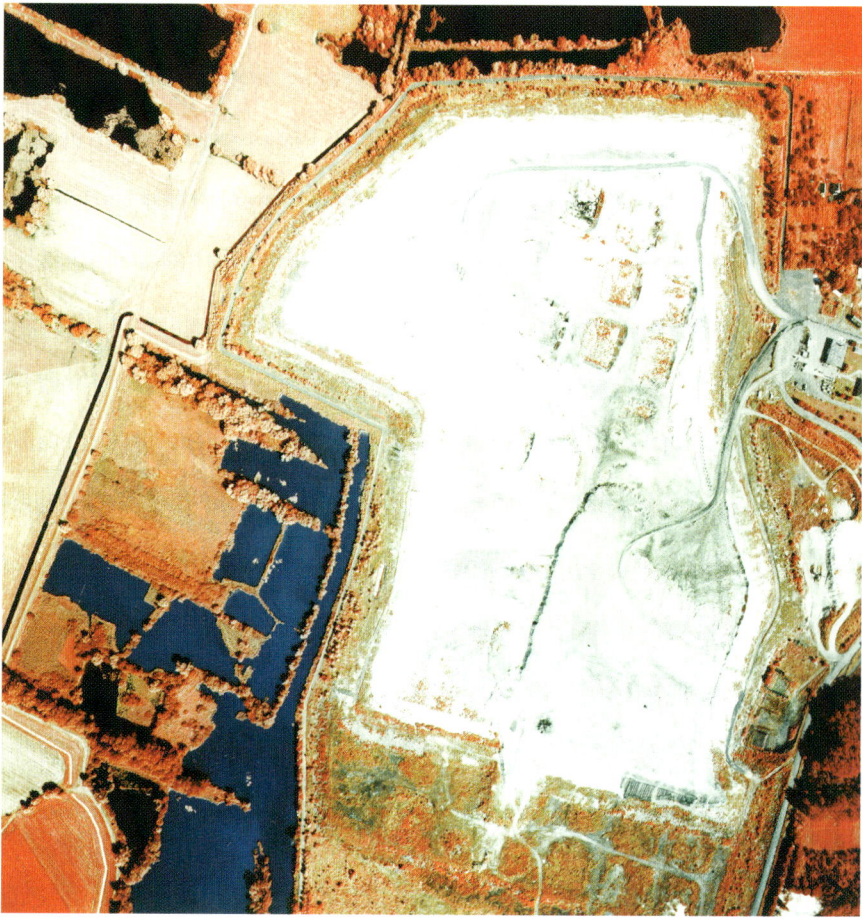

Fig. 3.4. CIR aerial photograph of the Vorketzin waste disposal site in Brandenburg, Germany; scale 1:5000. (taken July 28, 1990, by Berliner Spezialflug, Luftbild GmbH; printed by courtesy of the Gesellschaft für Umwelt und Wirtschaftgeologie mbH, Berlin)

Some geoenvironmental assessments may require oblique aerial photographs, particularly when overviews of large areas are needed. Oblique photographs may prove particularly valuable in the early phases of a project when the overall site is being characterized.

The oblique color photograph shown in Fig. 3.5 shows an area near the edge of an abandoned open-pit uranium mine at Lichtenberg (southwest of Ronneburg, Thuringia, Germany). This oblique photograph highlights the extent of environmental impact in the affected region. Old mine buildings, mine spoil and mill tailings, as well as tailings ponds for liquid wastes, can be seen. Although the impact on this region from decades of uranium mining can only be determined by a professional evaluation of vertical aerial photographs, the oblique image gives an

CHAPTER 3 · Obtaining Remote-sensing Data

Fig. 3.5. Oblique aerial photograph from an altitude of about 350 m showing the area of the former uranium open pit mine at Lichtenberg, near Ronneburg, Thuringia, Germany. Old mine buildings, mine dumps, and lagoons for liquid wastes are clearly visible (photo: F. Boeker, F. Kuehn, BGR, September 3, 1991)

overview and provides a basis on which to estimate the complexity of existing conditions and what clean up might entail (cf. Section 4.2.2).

Archival aerial photographs. In Germany, archives of aerial photographs are maintained by the German state survey offices, municipal administrative offices, commercial aerial photography firms, military institutions, private archives and users of aerial photographs. Specific sources are listed in Albertz (1991) and Strathmann (1993).

Any hazardous waste site characterization generally begins with an archival search for aerial photographs to reconstruct the historical development of the site. Land survey office archives are among the best sources because they store photo-

Fig. 3.6. American WW I aerial photographs showing the storage and shipment of military supplies at Andernach on the Rhine. Pits and waste dumps bordering this industrial area can be seen. These pictures are important for understanding the historical development of hazardous waste sites. Upper oblique photograph taken May 20, 1919, lower vertical photograph taken May 23, 1919 (courtesy of Luftbilddatenbank in Würzburg)

graphs from aerial surveys commissioned at regular intervals since the 1950s. The flights, usually repeated every five years, form the foundation for continuous updating of topographic maps. Availability, however, differs between the "old" and the "new" federal states of Germany (the states of the pre-reunification Federal Republic and the former Democratic Republic, respectively). Photographs from flights over the former East Germany from the 1950s to 1980s can be obtained from the Potsdam branch of the Federal Archives.

Many of the aerial photographs taken in the 1930s were made into aerial image maps and produced as topographical maps at a scale of 1 : 25 000 (known as *Reichsluftbildkarte*). These air-photo maps are available for large parts of Germany and can be obtained from the Federal Research Institute for Regional Geography and Regional Planning (Bundesanstalt für Landeskunde und Raumordnung) in Bonn. They may be used without any restrictions, but are not suitable for stereoscopic evaluation.

The archives of the American and British governments are additional sources of historical aerial photographs. Most wartime aerial photographs were taken during World War II; however, rarer photographs from WW I also exist (Fig. 3.6). In general, private consulting companies specialize in archive searches for these wartime aerial photographs.

During WW II, intensive aerial reconnaissance was conducted over Germany. These photographs were used to locate combat targets during the massive air raids (Fig. 3.7). Although aerial photographs taken as part of a land survey usually cover an extensive area, military photographs are generally taken on linear traces with sudden direction changes (Dech et al. 1991). For most wartime reconnaissance flights, Target Information Sheets exist which define the rationale for and the outcome of each bombing raid (Section 6.2). In general, it is possible to use these sheets to deduce which potentially toxic substances seeped into the soil after a raid on an ordnance plant or chemical factory. These sites should be considered potentially hazardous waste sites.

Mapping bomb craters is one of the "classic" evaluation tasks for which wartime aerial photographs are used. In the immediate postwar period, these craters often became repositories for ammunition and similar material, and may still be a threat because of chemicals or unexploded bombs.

These aerial photographs may also be used to locate abandoned waste disposal sites that were established on destroyed industrial and ordnance sites even those that are not recognizable from the ground or in recent remote sensing data. Wartime aerial photographs may be the sole source of information with which to assess the risks under these conditions. Figure 3.7 is a photograph of a destroyed industrial plant showing bomb craters in close proximity to a large pit (top center). Both of these depressions may have served as disposal sites for hazardous substances.

Post-war aerial photographs of Germany taken by the Allied Forces provide a link between the wartime photographs and land survey reconnaissance, which began in the mid-1950s. These images are available for all parts of Germany, and are primarily available through state and federal archives and private companies.

Fig. 3.7. WW II aerial photograph (March 24, 1945) of a destroyed industrial complex in the Dortmund area (Ruhr District). Damaged industrial plants, from which pollutants may have seeped into the soil, and bomb craters and pits, which may have served as dumps, are potentially hazardous sites (courtesy of Luftbilddatenbank in Würzburg)

3.2.2
Nonphotographic Imaging Sensors

3.2.2.1
Introduction

Instead of using light-sensitive films, nonphotographic imaging systems detect the incoming electromagnetic radiation with semiconductor detectors or special antennas. Although the arrangement of semiconductors in the focal plane of optical-electronic scanners is similar to aerial photographic cameras, optical-mechanical scanners and radar remote-sensing systems operate on entirely different principles.

Nonphotographic remote-sensing systems exhibit the following basic advantages over conventional aerial photographic cameras:

- the ability to detect reflected or emitted radiation from objects on the Earth's surface in the visible and longer wavelengths (near-infrared, thermal radiation, microwaves, and radar);
- the ability to directly record and store data in digitized form on tape/streamer or hard disk for later data processing;
- the ability to transmit data from the sensor to an Earth-based receiving station (providing immediate availability of the data);
- the ability to detect and measure EM radiation in narrow bands.

The general disadvantage of most mechanical scanning remote-sensing systems is their low spatial resolution. Thus, to obtain the most information, it is advisable to combine photographic and nonphotographic data. Processing and interpretation of nonphotographic data requires special software.

The theoretical background and technical principles of nonphotographic remote-sensing systems can be found in the literature. Overviews presented here are limited to those systems used for evaluating waste sites. For more detailed discussion (and for systems not discussed in this book, e.g., Lidar, microwave methods), see Colwell (1983), Kronberg (1985), Gupta (1991), Albertz (1991), Sabins (1996) and others.

3.2.2.2
Optical-Mechanical Line Scanners

Optical-mechanical line scanners produce images of the Earth's surface by scanning across a swath of land. In the literature, this scanner system is often referred to as a "whisk broom" scanner. Optical-mechanical line scanners consist of three parts (see Fig. 3.8):

- optical-mechanical line scanning system,
- detector unit,
- data recording system.

A mirror tilted at 45° is the basic element of an *optical-mechanical line scanning system*. It scans the area by rotating or oscillating perpendicular to the direction of flight. The size of a ground element resolved at a particular moment of the scanning process is defined by the **instantaneous field of view** (IFOV). The IFOV is given by the angle in radians of the field of view at a particular instant (Gupta 1991). Thus, the size of a the smallest discrete area on an image (pixel), depends on the flight altitude and the IFOV. The spatial resolution of most scanners is between 1 and 2.5 mrad. A scanner with an IFOV of 1.5 mrad, equivalent to 0.086°,

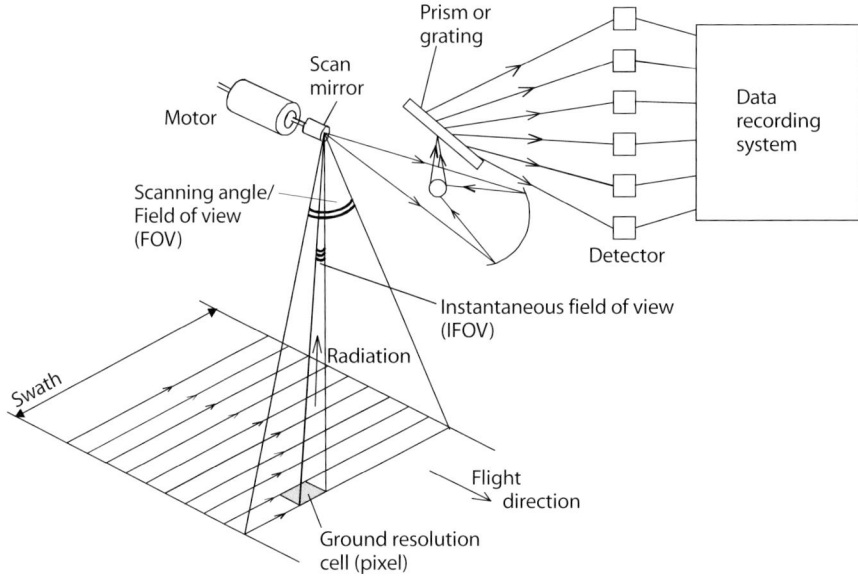

Fig. 3.8. The principle of optical-mechanical line scanners (modified after Gupta 1991)

will scan with a pixel size of 1.5 m × 1.5 m at a flight altitude (h_g) of 1 000 m. These data are for the center of a scanned swath. As a rule of thumb, an object on the Earth's surface should have a minimum size A of about 2.8 times the pixel size to be recognizable on a scanned image (Albertz 1991). For the preceding example (pixel: 1.5 m × 1.5 m; h_g = 1 000 m), an object should be at least 4.2 m.

When scanning flights are planned, there often has to be a compromise between spatial resolution and swath width (width of the strip of terrain surveyed). The margins of the scanned image usually show distortion of the pixels (panoramic distortion), which will later have to be geometrically corrected. Because of inherent aircraft instability, geometric rectification of aircraft data is a major problem when georeferenced data is required. With satellite-borne scanners, this distortion is normally negligible due to the orbital altitudes of the satellite.

A scan line is the result of a single scan across the width of the swath. The length of one scan line, i.e., the width of the swath, is defined by the angular field of view (FOV) of the scanner. This angle is usually between 90° and 120°. Due to the forward motion of the aircraft (or satellite), the terrain is scanned line by line and a continuous image is generated (Figs. 3.8 and 3.9). Only if the synchronization is not properly adjusted will overlaps or gaps appear between the scan lines. To produce high-quality images, perfect synchronization between the scan frequency of the mirror and the speed of the aircraft (or satellite) is crucial.

The optical signal from the ground reaches the scan mirror and passes through the optical system to the *detector unit*. The detectors transform the optical signals into electrical signals. The intensity of the electric signal depends largely on the intensity of the incoming radiation and, thus, on the spectral properties of the

Fig. 3.9. *Daedalus-AADS* (Table 3.2) scanner image of the Münchehagen waste disposal site in Lower Saxony, Germany. The image was taken on July 6, 1989, by DLR Oberpfaffenhofen for the BGR (*channel 3* = red, *4* = green, and *5* = blue)

ground surface. Multispectral scanners with up to 288 spectral channels have been developed, tested and applied to experimental problems. The radiation is divided by a system of gratings and prisms into partial spectra and then passed to specialized detectors. Normally, it is necessary to use several different detectors because different parts of the electromagnetic spectrum between the visible spectrum and thermal radiation require different, special semiconductor crystals. It is common to employ Si detectors for the visible range and NIR-I and PbS detectors for the NIR-II part of the electromagnetic spectrum. An InSb detector for thermal radiation in the MIR II range (Table 2.1) requires cooling with liquid nitrogen.

The electrical signals, commonly voltages, from the detectors are stored as digital data in the *data recording system* on magnetic tapes or other data storage media. In addition to the relatively low ground resolution, the major disadvantage of

Table 3.2. Spectral recording channels and band widths of the *Daedalus AADS-1268* airborne scanner (after DLR documents)

Channel number	Spectral range	Channel range (µm)
1	Visible light	0.420 – 0.450
2		0.450 – 0.520
3		0.520 – 0.600
4		0.605 – 0.625
5		0.630 – 0.690
6		0.695 – 0.750
7	Near infrared (I)	0.760 – 0.900
8		0.910 – 1.050
9	Near infrared (II)	1.55 – 1.75
10		2.08 – 2.35
11	Thermal infrared	8.5 – 13.0

optical-mechanical line scanners is their susceptibility to mechanical failure. Their many moving parts must be adjusted very precisely and are subject to wear. In addition, the previously mentioned distortion related to movements of the aircraft and panoramic distortion at the edges of the scanned swath, require computer corrections. Subsequent processing and interpretation of the data requires access to a digital image processing system.

To obtain optimum results, the geologic interpreter should be involved in the processing of the data so that he/she knows what has been done to the original data. Geographical and geologic knowledge about the target study area and a well-focussed, goal-oriented approach are prerequisites to a successful study.

Among the most widely used multispectral optical-mechanical airborne scanners are those of the *Daedalus* series (Table 3.2; Fig. 3.9). A more detailed explanation of the construction and performance of optical-mechanical scanners can be found in Kronberg (1985), Gupta (1991) or Sabins (1996).

The prime reason optical-mechanical line scanners are still in use, despite being technically outdated by optical-electronic scanners, is their ability to record thermal radiation from objects on the Earth's surface. In the wave of advanced technological development, it is assumed that future generations of optical-electronic scanners will be capable of routine data acquisition in the thermal wavelength ranges.

Waste sites do not usually cover areas larger than 1 km^2. Thus, highly sophisticated scanning systems designed for thermal mapping of larger areas may not be operated economically in every case. Simple thermographic imagers often fit the requirements of waste site characterization best.

The thermal images shown in Sections 4.2 and 6.3 were generated with an *AGEMA 900* thermography system. This system produces scanned images directly

Fig. 3.10. *AGEMA 900* thermography system prior to its installation in a *Cessna 206*. From left to right: scanner with mirror, Dewar flask for liquid nitrogen, computer with keyboard, monitor, and GPS

and can be operated from a small aircraft with a minimum of effort. However, other thermal systems such as *Inframetrics* and *NEC Thermo Tracer TH 1101* produce similar results.

The *AGEMA 900* scanner collects data in the 8–12 µm wavelength region and has a lens with an FOV = 20° and an IFOV of 1.5 mrad. During a flight, the computer may record more than 5 000 images on hard disk. Figure 3.10 shows the *AGEMA 900* system prior to its installation in a single-engine aircraft, a *Cessna 206*.

The thermal images can be processed and evaluated later with the same hardware and software used to record and store the data during the flight. The easy-to-use, software permits the generation of optimal thermal images using different processing techniques, such as enhancement of temperature patterns by stretching, plotting isotherms, determination of temperature along profiles and at point sites, and color coding of temperature intervals.

Figure 3.11 shows a thermal image of the western rim of the Schoeneicher Plan waste disposal site south of Berlin, taken with the *AGEMA* system. Unlike aerial photographs, the thermal imager can detect streams of air between the warm surface of the landfill and the colder surrounding area. Waste particles may be removed by local air streams and accumulate in the surrounding farmland.

More detailed information on options and limitations of thermal remote sensing for the investigation of waste disposal sites is given in Chapter 4 and Sections 6.3, and 6.4.

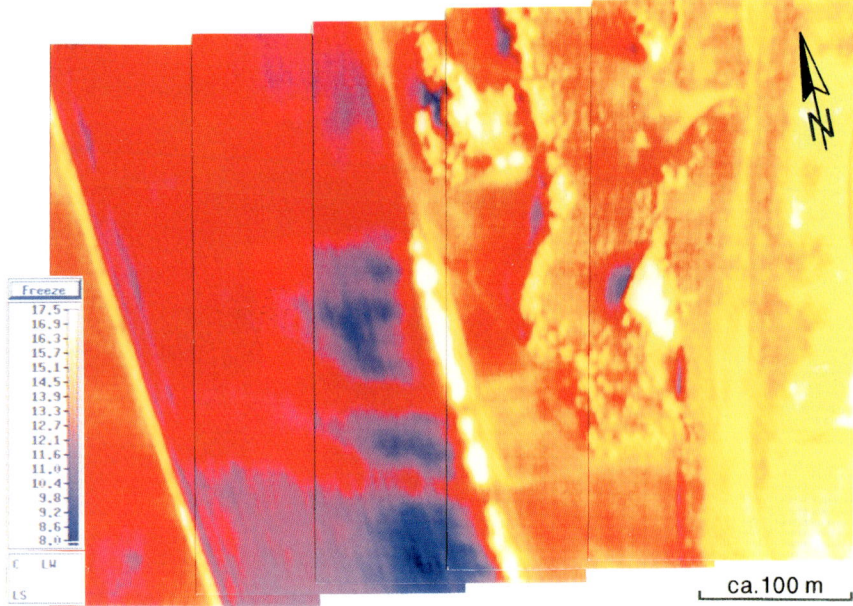

Fig. 3.11. Thermal image of the western rim of the Schoeneicher Plan waste disposal site and adjacent land. The margin of the deposit (*right*) is warmer than the field (*left*). Relief, stands of trees, and local winds generate exogenic thermal anomalies. Temperature in °C (image taken on May 13, 1993, at 05:19 CET by F. Boeker and F. Kuehn, BGR)

3.2.2.3
Optical-Electronic Scanners

Optical-electronic scanners are often referred to as CCD (**c**harge **c**oupled **d**evice) linear array scanners or push broom scanners. In principle, their construction is comparable to that of photographic cameras. Instead of a film, CCD arrays are installed in the focal plane of a camera-like recording system. CCD arrays consist of a large number of detector elements, e.g., silicon-based photo detectors. More than 1 000 detectors may be placed on a single 1.5 cm chip. Several of these chips are mounted in the focal plane of an optical-electronic scanner (see Fig. 3.12). The optical system focuses the incoming radiation on the surface of the light-sensitive chips. The radiation on each detector is transformed into electrical signals. The intensity of each individual signal depends on the "brightness" of the respective landscape feature. The detector chips are set up at right angles to the direction of flight (so that the swath is scanned in the manner a push broom is used).

The ground resolution cell (pixel) of an optical-electronic scanner depends on the size of the detector cell projected onto the ground surface. The ground resolution is, thus, dependent on the flight altitude h_g, the integration time τ of any single scan, and the velocity v of the aircraft. The ground resolution cell is described by a line L_1 parallel to the direction of flight (Eq. 3.4) and L_2 perpendicular to this direction (Eq. 3.5) as follows:

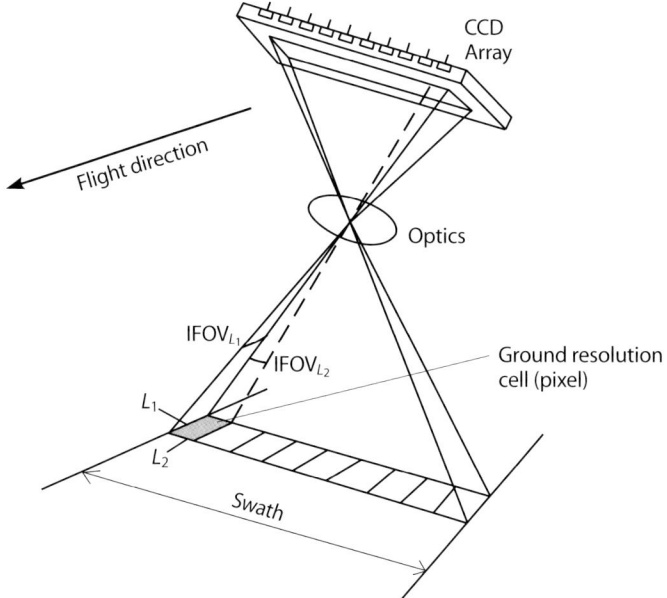

Fig. 3.12. Principle of an optical-electronic scanner (CCD linear array scanner, often called a push broom scanner) (modified after Gupta 1991)

$$L_1 = v\,\tau \tag{3.4}$$

$$L_2 = (a\,h_g)\,/\,c_f \quad, \tag{3.5}$$

where a is the distance between the centers of adjacent detector elements on a chip and c_f is the focal length of the optical system.

The major advantage optical-electronic scanners have over optical-mechanical scanners is their greater operating reliability, due to the lack of moving parts, higher geometric precision along a scan line, higher spectral resolution, and longer life-span. At present, it is still difficult to produce chips with rows of detectors that are sensitive in the thermal infrared, are close enough together, and can be individually cooled during flight. But it is only a question of time until the technology extends detection capability beyond the range of visible light (VIS) and parts of the near infrared (NIR-I and II) spectrum.

In multispectral optical-electronic scanners, the incoming radiation is divided into its spectral components by a grating or prism. By arranging several rows of detectors, it is possible to scan the surface over a selected wavelength interval. The *CASI* scanner (described in the following chapter) is a multispectral recording system based exclusively on the principle of an optical-electronic scanner.

Although infrequently used at present, optical-electronic airborne scanners have the potential of becoming one of the primary remote-sensing systems in the future.

3.2.2.4
Imaging Spectrometers

Imaging spectrometers, also called hyperspectral scanners, operate on the principles of the push broom and whisk broom. They record images of the Earth's surface in many narrow spectral channels in the visible and near-infrared parts of the electromagnetic spectrum. Thus, a continuous spectral signature for each pixel or ground resolution cell can be generated (cf. Gupta 1991, Sabins 1996). More detailed examples of imaging spectroscopy techniques and applications are given in Sections 6.6 and 6.7.

Several German remote-sensing companies use the Canadian airborne compact spectrographic imager *CASI* (Table 3.3 and Fig. 3.13). Theoretically, *CASI* uses 288 recording channels, each with a band width of 1.8 nm, but due to the large data volumes, data is normally acquired using considerably fewer channels. Further examples of the use of *CASI* data are included in the Sudbury mining district case study by Singhroy (Section 6.5).

The German Aerospace Center (DLR, Deutsches Zentrum für Luft- und Raumfahrt) has been using a digital airborne imaging spectrometer (*GER-DAIS-7915*) since 1995. This airborne spectrometer is capable of acquiring multispectral images and pixel-spectra using 79 channels in the wavelength range 0.4–12 µm. The band widths are 12–35 nm (spectral range 0.4–1.0 µm), 36–56 nm (1.5–1.8 µm), 20–40 nm (2.0–2.5 µm), 2000 nm (3.0–5.0 µm), and 600–1000 nm (8.7–12.3 µm).

NASA's hyperspectral *AVIRIS* scanner (*Airborne Visible InfraRed Imaging Spectrometer*) is used mainly by U.S. agencies and companies (see Sections 6.6 and 6.7, this volume). *AVIRIS* uses 224 spectral bands between 0.4 and 2.5 µm (IFOV of 1 mrad). Further hyperspectral instruments are the Italian *MIVIS* (Multispectral Infrared and Visible Imaging Spectrometer) (Bianci et al. 1997), and the Australian *HYMAP* (Hyperspectral Mapping) (no reference available, see www.intspec.com).

The use of these imaging systems will significantly improve the ability to directly determine the properties of materials on the Earth's surface. The work of Jansen (1994) illustrates the advantages of using imaging spectroscopy data in remote-sensing investigations. Jansen developed a computer program GenISIS

Table 3.3. Selected specifications of the *CASI* scanner (after documentation from WIB GmbH, Berlin)

Angular field of view (FOV)	35.4°
Number of detectors per scan line	512
IFOV	1.21 mrad
Spectral range	0.43 – 0.87 µm
Total number of spectral recording channels available	288 with 1.8 nm band width
Maxium number of channels that can be simultaneously used	15

RGB-true color combination
Channel 3: 663.9–680.5 nm = *red*; channel 2: 547.6–564.1 nm = *green*;
channel 1: 467.7–487.6 nm = *blue*

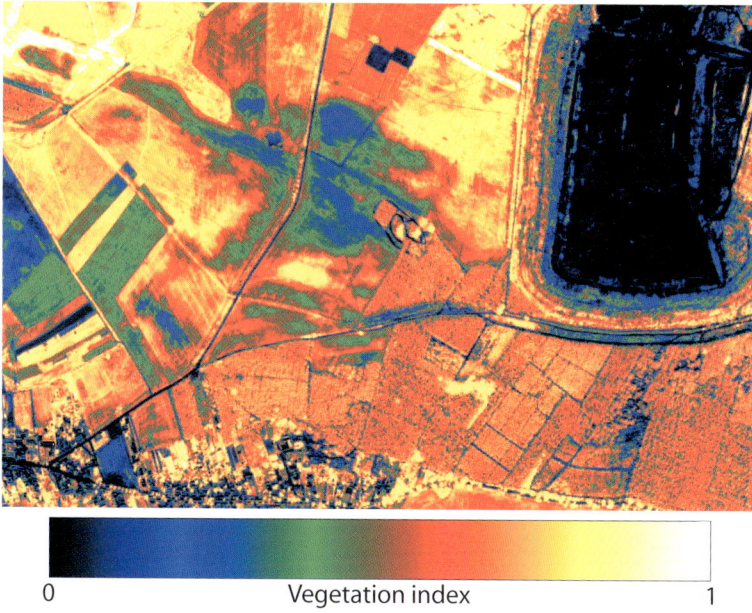

0   Vegetation index   1

Fig. 3.13. *CASI* image taken on June 26, 1993, showing the SW corner of the Schoeneiche waste disposal site (*right*) and the adjacent area. *Upper image*: RGB (*red/green/blue*) true color combination; *lower image*: the vegetation index $NDVI = (NIR - red) / (NIR + red)$, using one recording channel each in the red and near infrared (NIR-I) ranges of the spectrum. The green/blue zone in the *center* depicts poor growth on sandy, highly porous, soil; compare with the CIR image in Fig. 4.8 (images and image processing: WIB GmbH, Berlin, for the BGR)

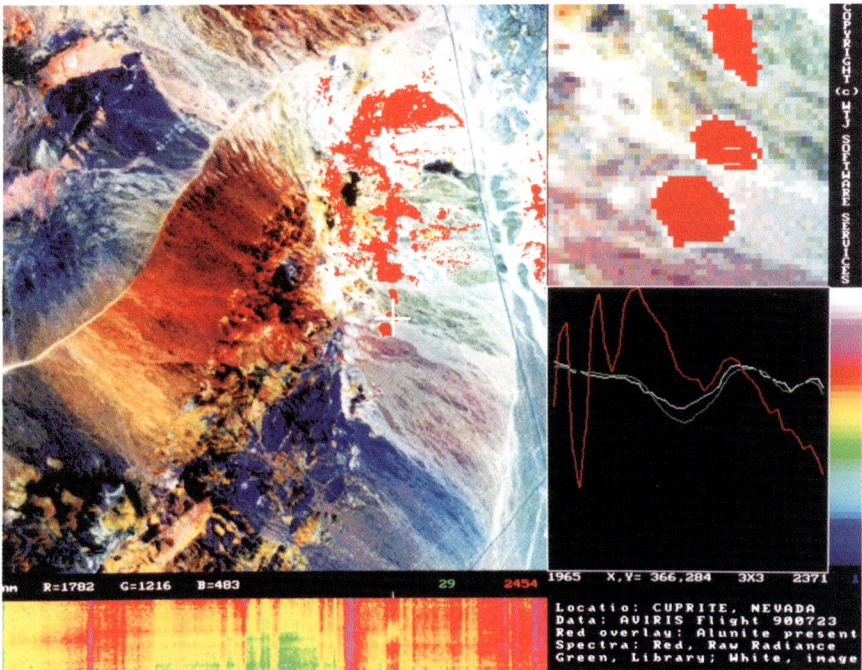

Fig. 3.14. Example of PC-based interpretation of data from an airborne imaging spectrometer; for explanation, see text (courtesy of W.T. Jansen, WTJ Software Services, San Mateo, California, USA)

(general imaging spectrometry interpretation system) for analyzing imaging spectroscopy data on the basis of comparison with spectral signatures in a database. Figure 3.14 shows an image of the Cuprite area of Nevada (USA) from *AVIRIS* instrument. Using only three of the 224 channels available over the ~400–2 450 nm wavelength range, Jansen produced a color-composite image of the hydrothermally altered rocks. The color-composite image (Fig. 3.14) was made on the basis of data from the channels for the wavelengths 1 782, 1 216, and 483 nm, and are depicted as red, green, and blue, respectively. The white cross marks an outcrop of alunite-rich rocks (red). The image on the upper right is an enlargement of this area. The spectral signatures of an extracted *AVIRIS* pixel and the spectrum from the database over the 1 965–2 371 nm wavelength range are shown on the lower right. Similarities between the spectra from the *AVIRIS* image (white) and the database (green) suggest the presence of alunite-rich rocks in all areas of the raw radiance image (red). For additional discussions on the acquisition, analysis, interpretation, and application of imaging spectroscopy (or hyperspectral) data, see the chapter on field checking (Chapter 5) and case studies by King et al. and Peters and Hauff (Sections 6.7 and 6.6, respectively). Imaging spectroscopy is well suited for geologic and geoenvironmental investigations.

3.2.2.5
Radar-Based Methods

In contrast to the previously discussed imaging systems, radar (*radio detection and ranging*) is an active imaging system. Radar "lights up" an area with its own radiation source in the cm-range of the electromagnetic spectrum (see Table 2.1) and is, thus, independent of natural radiation.

The comparatively long-wave electromagnetic radiation of radar penetrates clouds. This all-weather capability is one of the distinct advantages of the radar technique. Due to the oblique illumination angle of the transmitted radar signal to the Earth's surface, radar images are especially useful for detecting slight changes in relief due to the radar shadowing effect. This property is useful for mapping topographically expressed faults, (for example, see the Sudbury case study by Singhroy, Section 6.5). Radar also detects differences in soil moisture content. The disadvantage of radar data is its comparatively low resolution and the complexity of data acquisition and interpretation.

Radar is used primarily to survey large areas, particularly reconnaisance surveys; it is unlikely that radar will be used to routinely investigate waste disposal sites. However, interested readers can evaluate additional information in the literature (Colwell 1983; Trevett 1983; Kronberg 1985; Wooding 1988; Gupta 1991; Barrett and Curtis 1992; Sabins 1996).

Chapter 4

The Use of Remote Sensing in Waste Disposal Site Investigation

Friedrich Kuehn · Bernhard Hoerig

4.1
Investigative Objectives and Interpretative Criteria

Many old, but currently active, waste disposal sites were selected using criteria that do not comply with modern standards. Consequently, many of these sites require additional reevaluation and investigation to address the following site characteristics (Fig. 4.1):

- Characteristics of the lithology at the base of the landfill,
- Artificial or natural systems draining the site,
- Types of wastes and the landfill management,
- Potential impact and associated risks of contamination of soils and ground water in the site vicinity, and
- Potential for contamination of natural resources at and near the site.

In a site search for areas suitable for a new waste disposal site, a thorough site characterization, including geologic and hydrologic evaluation, should be conducted. For optimum characterization, remote-sensing methods should be included in the geologic and geophysical techniques used in the site characterization.

Aerial photographs are evaluated and interpreted using stereoscopic photograph pairs. Equipment used for interpretation ranges from simple mirror stereoscopes to sophisticated stereoplotters with computer-based photogrammetric evaluation and mapping programs (see Albertz 1991).

As discussed previously, there is a distinction between thematic and photogrammetric evaluation of aerial photographs. *Photogrammetric interpretation* of aerial photographs requires the determination of the three-dimensional coordinates (x,y,z) of ground control points, buildings, etc. This technique allows highly accurate topographic maps to be produced. In contrast, geologic and photogeologic maps, maps of suspected hazardous waste sites, and hydrological maps are created by *thematic interpretation* of aerial photographs, which also requires a high degree of accuracy in topographical classification.

The most important image characteristics and landscape factors used in mapping soils, rocks, and geologic structures from aerial photographs and digital imagery are the *photographic gray scales*, *morphology*, *vegetation*, *drainage systems*, *geologic structure*, and *patterns related to land use*. These features are described, discussed and illustrated by Kronberg (1984) and Sabins (1996).

Environmental evaluation using remote-sensing data requires examination of additional criteria. Indicators of tectonic fractures and faults, potential pathways for pollutant migration, and the permeability of the subsoil layers, as well as *anthropogenic features* of a landscape, need to be considered. The features that best identify anthropogenic components of a landscape in aerial photographs include the following (Dodt et al. 1987):

- *Anomalous relief* that may indicate landfill, waste-disposal sites, mine spoil heaps, excavation work, etc.
- *Anomalous gray-scale patterns* that might indicate contaminated soils, filled-in depressions, spread of leachate from waste disposal sites, abandoned industrial sites, etc.
- *Vegetation anomalies* that might indicate contaminated soils and ground water.

The *time* of acquisition of the aerial photography is a critical factor in maximizing the utility of aerial photographs for interpretation (Boeker and Kuehn 1992). Because vegetative cover often masks geologic features on aerial photographs, photographs for geologic analysis should be taken during seasons with a minimum of vegetation. Experience has shown that aerial photographs taken after the snow has melted and the surface has dried, but before the vegetation has fully emerged, are well-suited for detecting faults, fractures, and channel systems. On the other hand, the distribution of soil types, lithologic contacts, tectonic structures, and the existence of pollutants in specific areas may be recognized by specific types and patterns of vegetation. Consequently, the experience of the remote-sensing specialist is crucial in determining the appropriate strategy for acquisition and interpretation of aerial photographs.

An assessment of geologic and environmental factors requires different approach from those needed for compilation of topographic maps. Evaluation of aerial photographs for geologic and environmental applications require consideration of various additional factors. For example, weather conditions are an important factor when acquiring images for geologic interpretation, because the amount of moisture in the soil can greatly affect the usefulness of the data. In some instances, it will be useful to acquire data with the sun at a low angle ($10°-30°$), because low-relief forms are enhanced by the more pronounced shadows on the ground, thus increasing the chance of detecting subtle geologic features, such as faults and fractures, as well as landfills.

Multitemporal analysis uses images or photographs taken over a period of time to detect changes, and is useful for environmental investigations.

Private companies and governmental agencies specializing in the evaluation of aerial photographs require state-of-the-art digital mapping and systems for evaluating and interpreting aerial photographs. These systems consist of high-quality stereoscopes with zoom optics, as well as hardware and software to measure image coordinates and parallax for precise determination of distances and terrain elevations. Integrated computer systems allow mapping of terrain features and storing the data in a GIS database.

If such, usually complex and comparatively expensive equipment is not available, a rather simple approach can be taken for an *interpretive synoptic evalua-*

tion of an area. The minimum requirement is a simple mirror stereoscope; however, the interpretation will be less precise. When a mirror stereoscope is used, significant terrain features can be marked on a transparent sheet placed over the photo. This sketch may then be brought to the desired scale with as little distortion as possible using a optical pantograph, or similar equipment and then transferred to the map. Distances can be measured using the scale of the aerial photograph or map, and elevations H_o can be estimated by measuring the length l of a shadow cast by an object, the image scale constant m_b, and the altitude of the sun's position h_s in degrees:

$$H_o = l \, m_b \tan h_s \tag{4.1}$$

Information on the composition of surface materials and identification of anomalous areas in a landfill can be obtained from thermal images. The properties of an object on the Earth's surface, as seen in a thermal image, are determined both by the component material and by the character of the surrounding environment. Similarly, the intensity of radiation in the thermal infrared range is dependent on both the surface temperature and the emissivity of the materials. Radiation temperatures calculated using measured radiation intensities generally correlate with temperatures measured at the terrain surface. Thus, calculation of surface temperatures requires considerable field work and is necessary only for special applications. The following physical and environmental parameters influence the thermal behavior of an object at the Earth's surface (Shilin 1980; Kronberg 1985):

- Physical parameters
  - Color
  - Composition
  - Surface properties
  - Density
  - Porosity or pore volume
  - Permeability
  - Water content
- Environmental parameters
  - Topographical position in the field
  - Orientation of the surface to the sun
  - Meteorological conditions
  - Microclimate
  - Humidity
  - Time of day, season
  - Type and extent of vegetation cover

Consequently, any given object in a thermal image can be identified as "warm" or "cold" relative to its surroundings. Environmental influences can cause confusing overlap or even mask thermal radiation emitted from natural or artificial objects, creating false anomalies in the thermal images. The diurnal temperature cycles of two very different materials show that it is impossible to generalize about the "warm" or "cold" character of materials on the basis of their physical characteristics. According to Gebhardt (1981), physical and environmental factors may act independently of each other or they may be synergetic or antagonisticantoa. In addition, the environmental factors are difficult or impossible to determine only on the basis of the images. A more extensive discussion of the possibilities and limitations of using thermal images for evaluating waste disposal sites is given in the section describing the Schoeneiche disposal site (Section 6.3).

We will use selected sites to illustrate the thematic interpretation of remote-sensing data for investigation of waste disposal sites and the adjacent areas, and subsurface conditions. The possibilities and limitations for the characterization of a site are dependent on the landscape and geologic conditions. These examples provide criteria for selecting appropriate remote-sensing methods for environmental site-characterization.

The investigative goals and evaluation criteria discussed above are typical for Germany, where distances are comparatively short (<1–5 km) between disposal sites, towns and villages, ground water protection zones, cultivated land and dense population centers. We have found that advanced multispectral approaches are often not the best means for disposal site investigations given the outlined conditions.

Entirely different conditions prevail in Canada and the United States, as presented in the case studies by Singhroy (Section 6.5), Peters and Hauff (Section 6.6), and King et al. (Section 6.7). Unlike Germany, low population densities predominate and the study areas may have retained some of their natural character. In addition, the study areas are large; the migration path of the contamination can exceed 50 km. As a result, the most appropriate methods for remote-sensing site characterization are different from those used in Germany. In some instances, high-resolution multispectral data are needed for characterization, and in other circumstances lower resolution satellite data or aerial photography are most appropriate (see Sections 6.5, 6.6 and 6.7).

The spectral ranges and imaging systems appropriate for addressing specific problems will be discussed in detail in the following discussion of waste disposal site-characterization, for which a generalized representation is shown in Fig. 4.1. The spectral ranges and remote-sensing systems that have the best potential for yielding the best results for a given task are indicated after the heading for each example. If several different spectral ranges, imaging systems, or film types provide similarly valuable information, then the most cost-efficient combination (spectral range/film/system) is shown in *italics*. The following abbreviations are used: AP: aerial photo; BW: black-and-white; CIR: color infrared; MS: multispectral; and VIS, NIR, and MIR as defined in Table 2.1.

Fig. 4.1. Generalized landfill with permeable base and without drainage water collection system (*top* is a top view and *bottom* is a cross section) with evidence of structures, characteristics, and properties that may be identified with remote-sensing techniques, depending on the site conditions (examples in Figs. 4.2–4.17; legend on facing page)

Top View

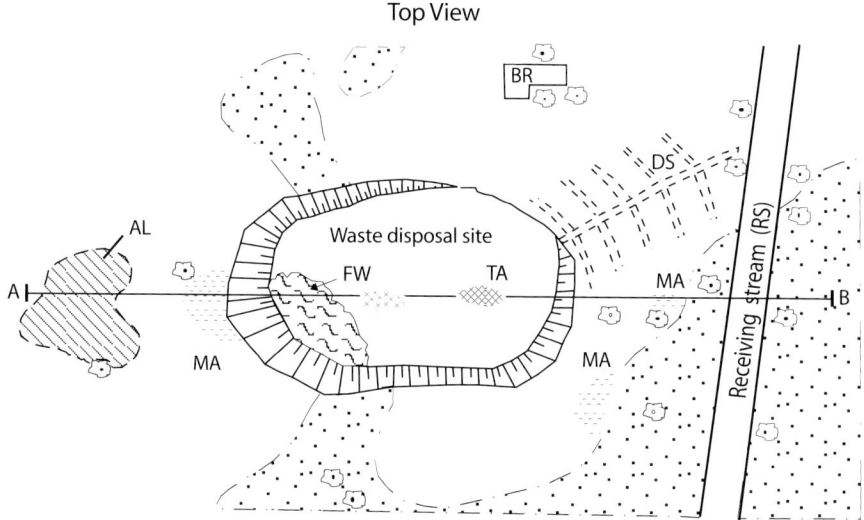

Cross-section A–B

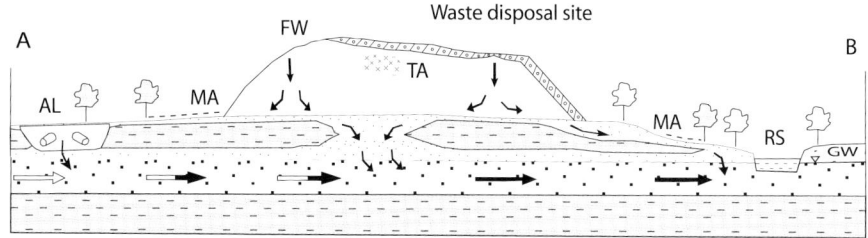

Legend

▱	Impermeable sediments	▱	Fresh waste (FW)
▱	Permeable sediments	▱	Covering of waste disposal site, e.g., loam stratum
▱	Gaps or weak points in the cover of the waste disposal site	▱	Building / settlement remains (BR)
▱	Wet areas (MA), escape of leachate	▱	Tree (uncontaminated), no root contact with contaminated ground water
▱	Receiving stream ditch (RS)	▱	Tree (contaminated), root contact with contaminated ground water
▱	Drainage system (DS)		Ground Water:
▱	Leachate	⇒	Clean ground water
▱	Thermal anomaly (TA)	→	Precontaminated ground water
▱	Abandoned landfill (AL)	→	Contaminated ground water
		▽GW	Groundwater surface of uppermost aquifer

4.2
Case Studies

4.2.1
Characterization of a Waste Disposal Site

Chronological Development of a Landfill Operation, including the Feasible Handling and Quantities of Dumped Wastes

Spectral range (type of data): *VIS (BW, CIR, Color AP)*, NIR-I (IR AP), MS (scanner)

Figures 4.2 and 4.3 are aerial photographs from 1992 and 1967, respectively, of the Schoeneicher Plan waste disposal site. An interpretation of the 1967 photograph by Krenz (1991) is included in Fig. 4.3. These sequential aerial photographs allow multitemporal investigation of the disposal site. If the aerial photographs are of good quality and their scale is sufficiently large, areas in which waste was dumped during earlier phases of site operations can be identified and potential risks evaluated.

Fig. 4.2. Black-and-white aerial photograph (*photomosaic*) showing the areal extent of the Schoeneicher Plan waste disposal site on May 15, 1992 (source: Landesvermessungsamt Brandenburg)

Fig. 4.3. Aerial photograph taken June 23, 1967 (**a**), and interpretation by Krenz (1991) on topographic base map from 1982 (**b**) (source: Federal Archives, Potsdam branch)

CHAPTER 4 · The Use of Remote Sensing in Waste Disposal Site Investigation

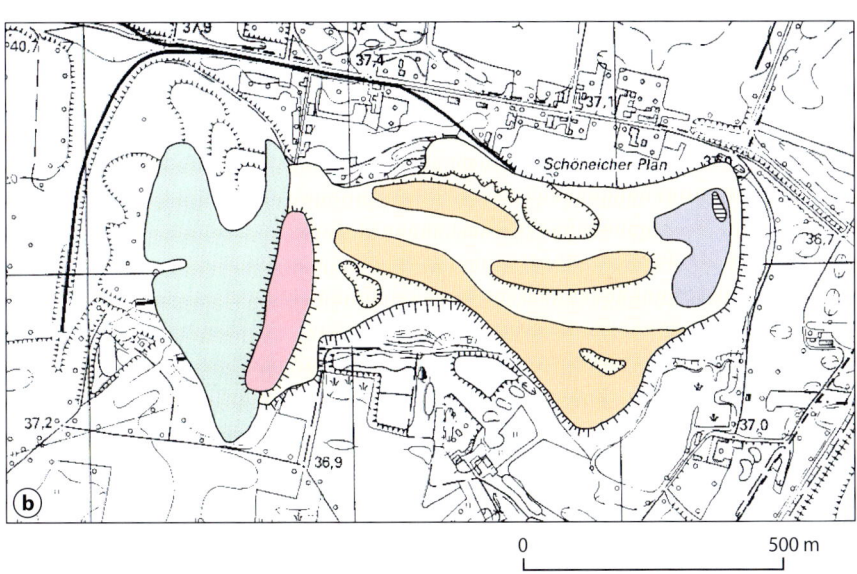

Legend

Topography after topographical survey sheet TK 10

- Abandoned landfill, covered, partly overgrown
- Abandoned landfill, uncovered
- Recent landfill
- Area for liquid waste
- Open water surface
- Slope between the abandoned and the recent landfill

Localizing Heat Sources Inside a Waste Heap

Spectral range (type of data): MIR-II (scanner)

Mining activities commonly result in changes to the original site. Investigations are necessary when risks are identified that threaten people, buildings and natural resources. The detailed black-and-white aerial photograph in Fig. 4.4a depicts parts of an abandoned open-pit mine in Thuringia, Germany. Mine dumps, mill tailings, tailing ponds, and abandoned mining equipment, are visible (Fig. 3.5). The

Fig. 4.4 a. Aerial photo taken April 24, 1995, showing a mining landscape in Thuringia, Germany. The dominant landscape structures include part of an abandoned open-pit mine (being filled with tailings), waste heaps, and an abandoned smelting plant (source: Landesvermessungsamt Thüringen)

mill tailings contain smoldering pyrite and carbon-rich schist. The smoldering is sustained by oxygen seeping through fissures and crevasses in the heap. This may lead to contamination of the ground water if rainwater percolates through the heap (Fig. 4.4c).

The thermal scanner image in Fig. 4.4b, taken at night, includes the area visible in the black-and-white aerial photograph in Fig. 4.4a. Areas of high temperature in the heap are indicated in the thermal image although the smoldering material lies several meters deep in the heap. The range of apparent surface temperatures (in °C) depicted in the image is indicated by the scale in the upper left corner. The yellow color denotes temperatures between 25 and 35 °C, the maximum. Monitoring of waste heaps with thermal systems provides information on the temporal and spatial extent of the oxidation zone. Rainwater percolating through smoldering pyrite and carbon-rich schist will form sulfuric acid, which will contaminate the ground water. These smoldering zones must be identified for effective remediation. The situation and the method is explained in Fig. 4.4c.

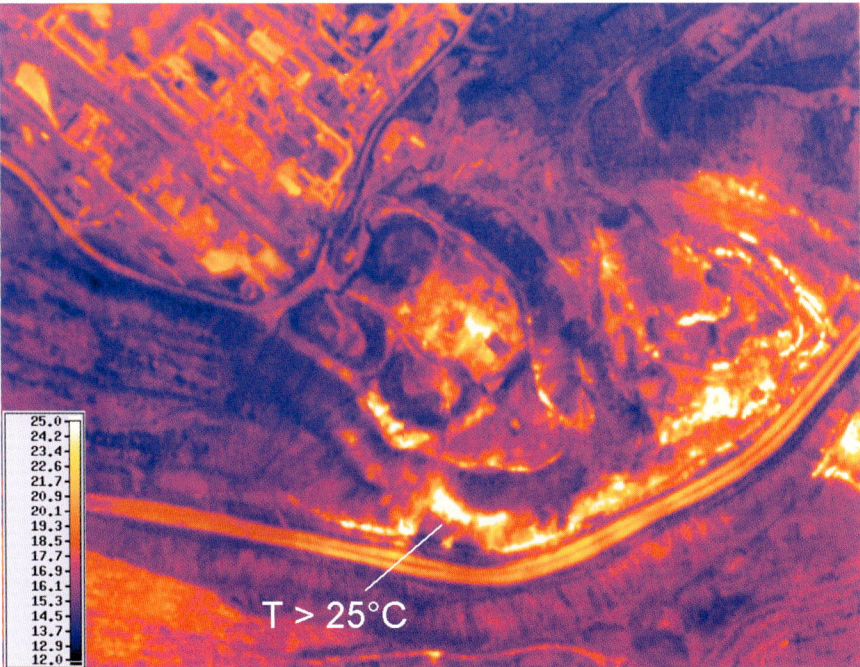

Fig. 4.4 b. Thermal image taken at 03:45 a.m. on May 23, 1994, of the abandoned mine site shown in Fig. 4.4a. Temperature anomalies (*bright yellow*) outline areas of smoldering pyrite and carbon-rich schist measured through several meters of cover. These zones of buried smoldering material are marked by significant temperature anomalies at the ground surface (see Fig. 4.4c; image: F. Kuehn, BGR)

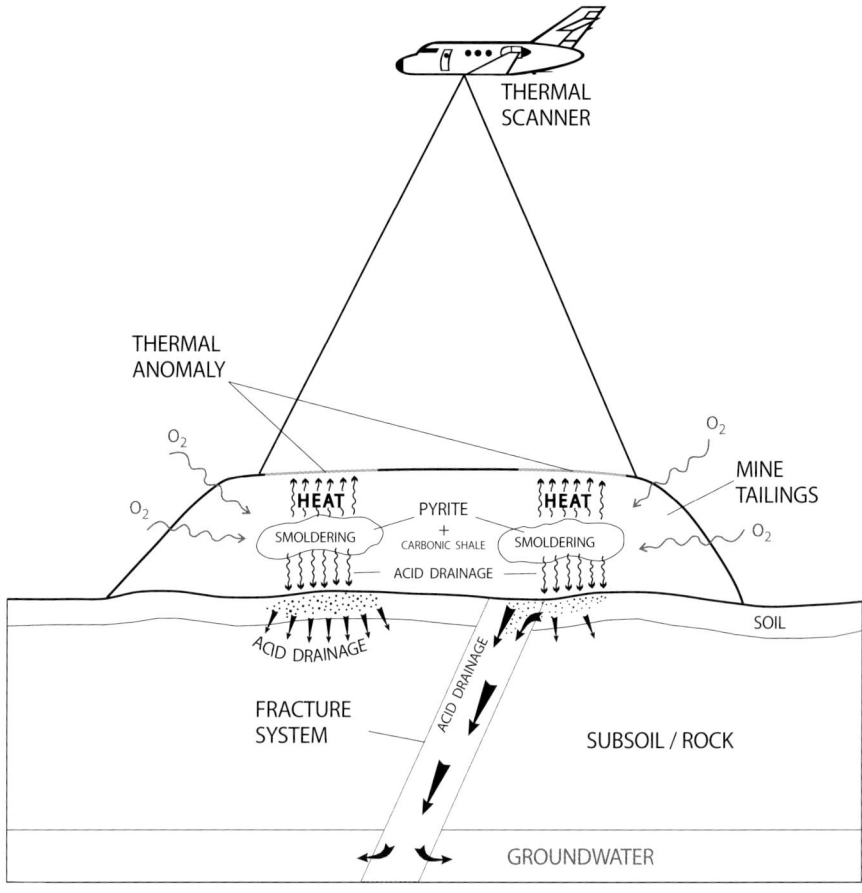

Fig. 4.4 c. Smoldering in pyrite and carbon-rich schist generates heat within the mine waste heap. Smoldering is sustained by an influx of oxygen through fissures and crevasses in the heap. Sulfuric acid is produced, which may lead to ground water contamination. The buried smoldering zones are easily detectable by thermal remote sensing

Seepage of Water at the Edges of Waste Disposal Sites

Spectral range (type of data): *VIS (BW, CIR, Color AP)* NIR-I (IR AP),
MIR-II (scanner), MS (scanner)

Anomalous plant growth, discoloration of the soil, or visibly wet areas at the edge of a landfill indicate seepage of water from the landfill and potential ground water contamination at the site (Fig. 4.5). In general, water that seeps from of a landfill is likely to be highly contaminated. Once leachate from a landfill is identified, further action is needed to characterize the leachate and curtail its movement.

CHAPTER 4 · The Use of Remote Sensing in Waste Disposal Site Investigation   43

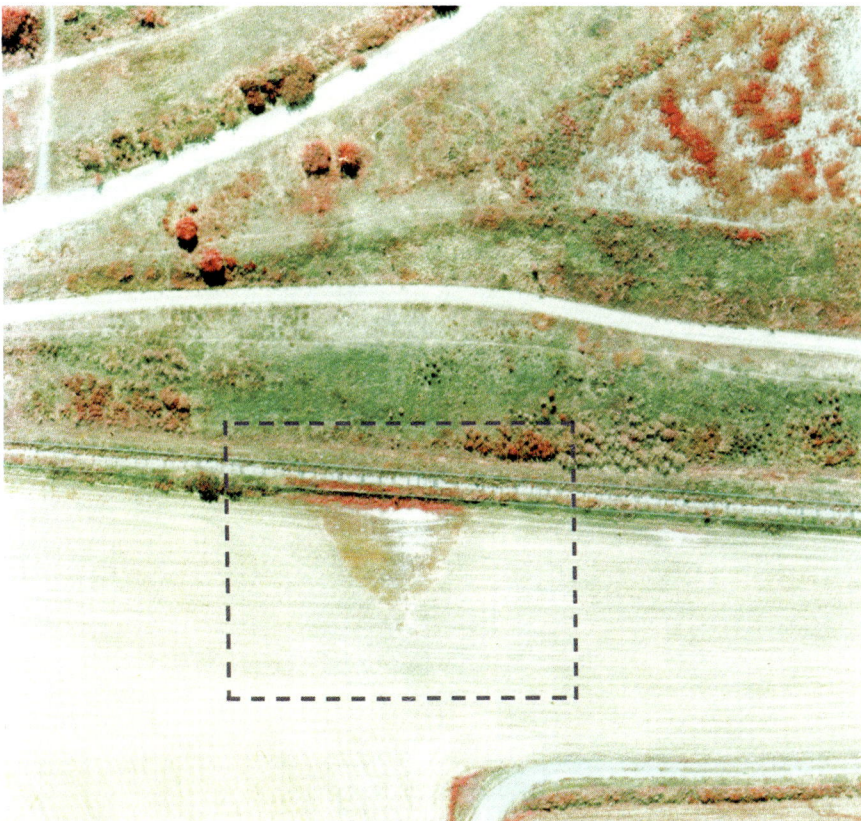

Fig. 4.5. CIR aerial photograph taken August 2, 1990, showing signs of water leachate (within the *dashed line*) near the edge of an abandoned disposal site that was transformed into a park in Berlin (aerial photograph: Eurosense GmbH, printed courtesy of the Senat für Bau- und Wohnungswesen, Berlin)

Succession of Vegetation on a Landfill containing Industrial Wastes

Spectral range (type of data): *VIS* (BW, *CIR*, Color *AP*), NIR-I (IR AP), MS (scanner)

The CIR aerial photograph in Fig. 4.6 shows part of the mine heap from an openpit lignite mine at Bruckdorf in Sachsen-Anhalt, Germany. The mine heap consists of Quaternary and Tertiary sediments and lignite particles, and was used over a long period of time as an industrial dump site. The natural succession of vegetation in the area is related to site conditions (soil cover, water supply, and nutrient supply) and the influence of the industrial wastes. The parts of the site that have a salt efflorescence crust remain unvegetated. Vegetation with poor vitality is visible in the CIR photograph along the edges of the site. The

Fig. 4.6. CIR aerial photograph taken August 22, 1991, of the dump of the former open-pit lignite mine at Bruckdorf (Sachsen-Anhalt, Germany) where industrial waste has affected the natural vegetation (succession). Salt efflorescence (*SA*) and seeping water (*SW*) can be recognized (photo: Hansa Luftbild; printed courtesy of Umweltamt Halle, in collaboration with C. Glaeser, Martin-Luther-Universität Halle, Germany)

type and degree of damage to the vegetation can contribute to the characterization of dumping areas and identifying the boundaries between different types of buried wastes.

4.2.2
Exploring the Immediate Vicinity Around a Waste Disposal Site

Springs and Soil Moisture Anomalies

Spectral range (type of data): *VIS* (*BW, CIR, Color AP*) NIR-I (IR AP), MIR-II (*scanner*), MS (scanner)

Springs and moist areas associated with a lack of ground cover are indicated by dark gray in black-and-white aerial photographs. In CIR images, areas of high moisture content can also be recognized by the associated vegetation shown as intense reddish colors. The ability to detect moist areas in thermal images is a function of temperature at the ground surface, the season, the type of ground cover,

CHAPTER 4 · The Use of Remote Sensing in Waste Disposal Site Investigation   45

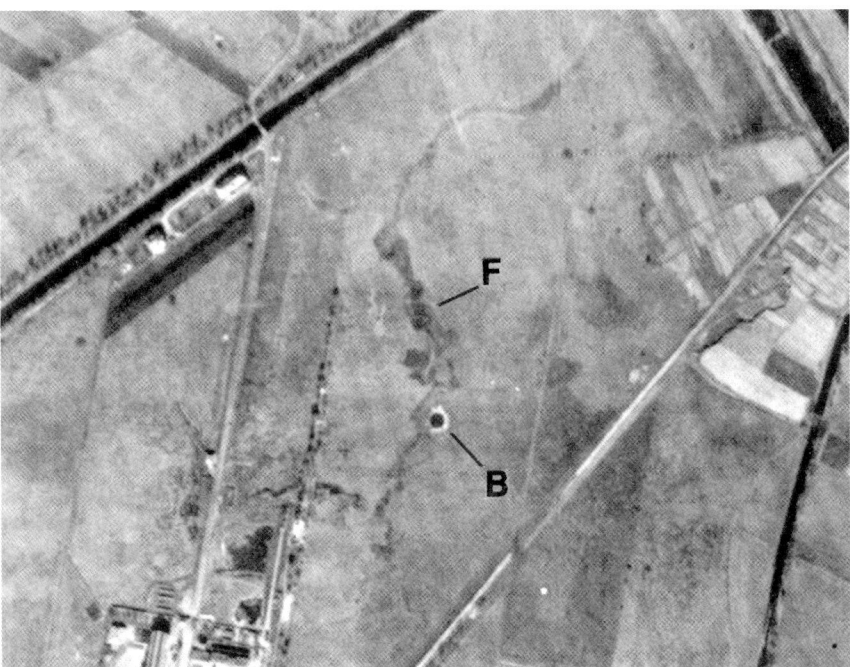

Fig. 4.7. Black-and-white aerial photograph taken April 10, 1945, of the terrain north of the Schoeneiche landfill near Berlin, Germany. Dark gray indicates areas with a relatively high soil moisture content. A bomb crater (*B*) can be seen on a distinctly recognizable silted-up stream (*F*); see also Fig. 3.3 (archives research and acquisition by the firm Luftbilddatenbank in Würzburg, Germany, for BGR)

and the time of day (pre-dawn is best) (see Section 6.3.2.3). Signs of wetness on the surface may be associated with slight changes in topography or the presence of impermeable, near-surface beds. If wet areas are detected near a waste disposal site, such as in Fig. 4.7, analysis of surface and ground water sampling is recommended. The results of these analyses help to determine the pathways for the migration of contaminated leachate from the waste disposal site and identify potential hazards.

Distribution of Areas of High and Low Permeability

Spectral range (type of data): *VIS (BW, CIR AP)*, MIR-II (scanner), MS (scanner)

Plant growth depends, among other factors, on the permeability of the soil and the underlying beds. The lighter colored areas in Fig. 4.8 correlate with areas of poor plant growth due to the sandy soil and the permeable beds there. Precipitation and leachate from the neighboring landfill can migrate through this permeable material directly into the aquifer.

Fig. 4.8. CIR aerial photograph taken July 3, 1993, of an area west of the Schoeneiche waste disposal site. Extensive sandy soil substrates, recognizable by the lighter colors, which has led to impaired plant growth (aerial photograph: WIB GmbH for the BGR)

Natural and Artificial Drainage Systems

Spectral range (type of data): *VIS* (*BW, CIR,* Color AP) NIR-I (IR AP),
  MIR-II (*scanner*), MS (scanner):

Natural surface and subsurface channels can act as pathways for leachate from landfills that do not have a basal liner. Subsurface pathways may result from many conditions, including frost-induced fissures and crevasses.

The features seen within the dashed lines in Fig. 4.9 are surface expressions of fissures caused by frost and which have been filled with permeable material (Merkt and Boeker 1993). The yellowish colored areas indicate soil with impaired vegetation resulting from the permeable fill of the fissures.

An artificial drainage system may have to be installed when water flow is blocked by impermeable rocks and soils or if the water table is high. A flat lowland adjacent to a waste disposal site is shown in a conventional black-and-white aerial photo-

Fig. 4.9. Oblique aerial photographs showing structures caused by frost action (frost polygons) in glacial till (within the *dashed line*). The frost cracks are filled with sandy, permeable material. At waste disposal sites such polygonal patterns of fissures can be ideal systems for migration of leachate from the landfill (courtesy of J. Merkt, NLfB, Hannover, Germany)

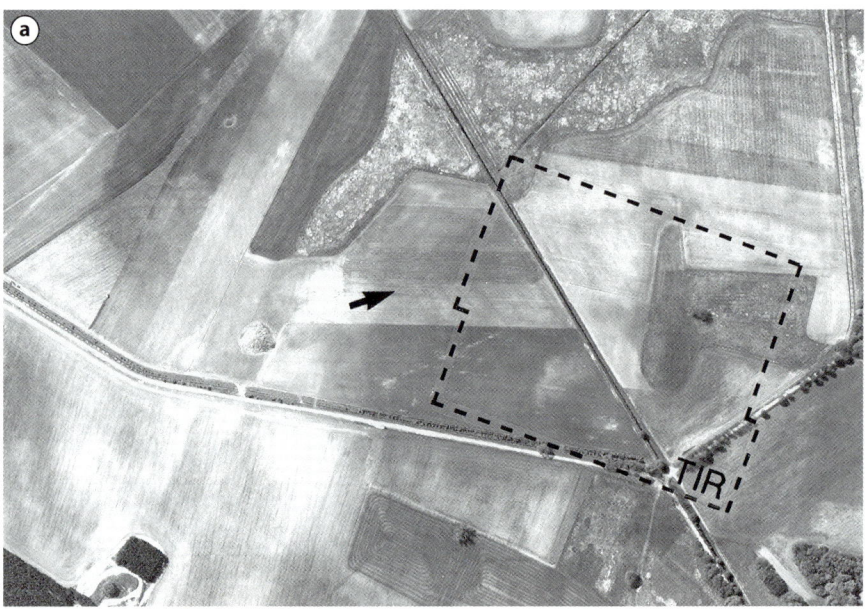
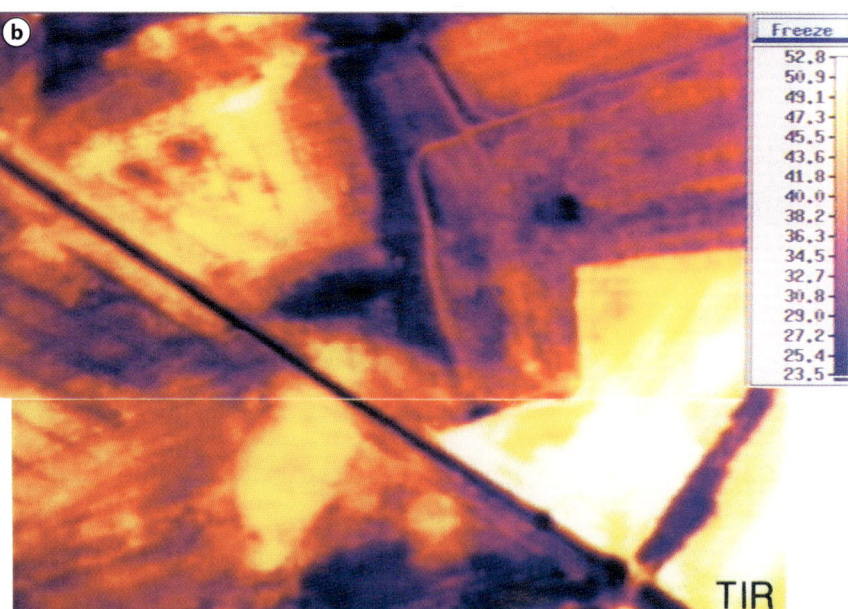

Fig. 4.10. Black-and-white aerial photograph taken May 15, 1992 (**a**), and a daytime thermal image, MIR-II (TIR), taken May 11, 1993 (**b**). The herring-bone patterns of an artificial drainage system are weakly visible in the black-and-white aerial photograph (*arrow*). This pattern can be more clearly seen in the thermal image. Where the drainage system still functions, the soil is relatively dry and "warm" (aerial photo: Landesvermessungsamt Brandenburg; thermal image: F. Boeker and F. Kuehn, BGR)

graph in Fig. 4.10a. Geologically, this area has Holocene to Pleistocene humic sands, muds, and bog lime deposits. The aerial photograph was taken at the beginning of a period of intensive plant growth and relatively high soil moisture content (mid-May). Features that may indicate an artificial drainage system (*arrow*) can be faintly seen.

The thermal image (MIR-II) in Fig. 4.10b, taken under comparable conditions, depicts the surface drainage more clearly because the drained ground has a higher temperature than the surrounding, moister soil. In this case, the intensity of temperature anomalies can be used as an indicator of the efficiency of the drainage system.

If a waste disposal site shows signs of artificial or natural drainage systems on or near the site, the site should be inspected to determine whether leachate from the landfill is drained by them (see Section 6.3).

Fractures and Faults

Spectral range (type of data): *VIS* (*BW*, *CIR*, *Color AP*) NIR-I (IR AP), MS (scanner)

Disturbed ground and fractures are often easily detectable in aerial photographs (Fig. 4.11). The surface expression of such features is marked by changes in relief, drainage patterns, vegetation growth patterns, and rock and soil color differences. The surface traces of fractures and faults are straight, which generally makes them easier to recognize in aerial photographs and other remote-sensing images. If topographically expressed, vertical faults and fracture systems can be best observed in images obtained at low sun angles.

Fractures in rock enhance weathering and erosion, thus promoting migration of leachate. Fractures detected at waste disposal sites are possible migration pathways and natural drainage systems that allow the outflow of contaminated water. The outflow is presumably along the fractures. Where wells are located along the strike of the fractures, the ground water should be analyzed for possible contamination (see the example of the Arnstadt/Eulenberg waste disposal site in Section 6.2).

Destabilization of Landfill Slopes

Spectral range (type of data): *VIS* (*BW*, CIR, Color *AP*)

Under certain geologic conditions, not only is there potential risk for soil and ground water contamination, but the stability of waste disposal sites and heaps may be threatened. The stereo-pair of aerial photographs in Fig. 4.12 shows part of a tailings heap south of Magdeburg, Germany. This heap (the white area on the left) contains waste from potash production. A sinkhole caused by collapse in the potash mine appeared near the heap in 1975 (Loeffler 1962; Brueckner et al. 1983). Stable conditions apparently prevail on the opposite side of the sinkhole from the heap (no concentric collapse fractures). However, the terrain surface is gradually collapsing along concentric fractures in the direction of the heap. Aerial photographs taken over a period of time can be interpreted to determine the stability of the area and predict additional collapse (Kuehn et al. 1997, 1999). If subsidence

Fig. 4.11. Oblique aerial photograph taken June 17, 1989, showing the moisture accumulating effect of vertical silt-filled joints in a normally dry Triassic limestone (Muschelkalk). Broad streaks in the foreground are traces of periglacial dorrs that, like the orthogonal joints on the other side of the road, have a fine-grained fill. If these structures occur at a waste disposal site, leachate from the landfill can migrate through them into the underlying aquifer, depending upon the permeability of the fill (photo: F. Boeker)

continues, additional geotechnical techniques should be used to monitor progressive deformation and erosion of the sediment layers at the base of the heap that might allow water to seep from the interior of the heap.

CHAPTER 4 · The Use of Remote Sensing in Waste Disposal Site Investigation 51

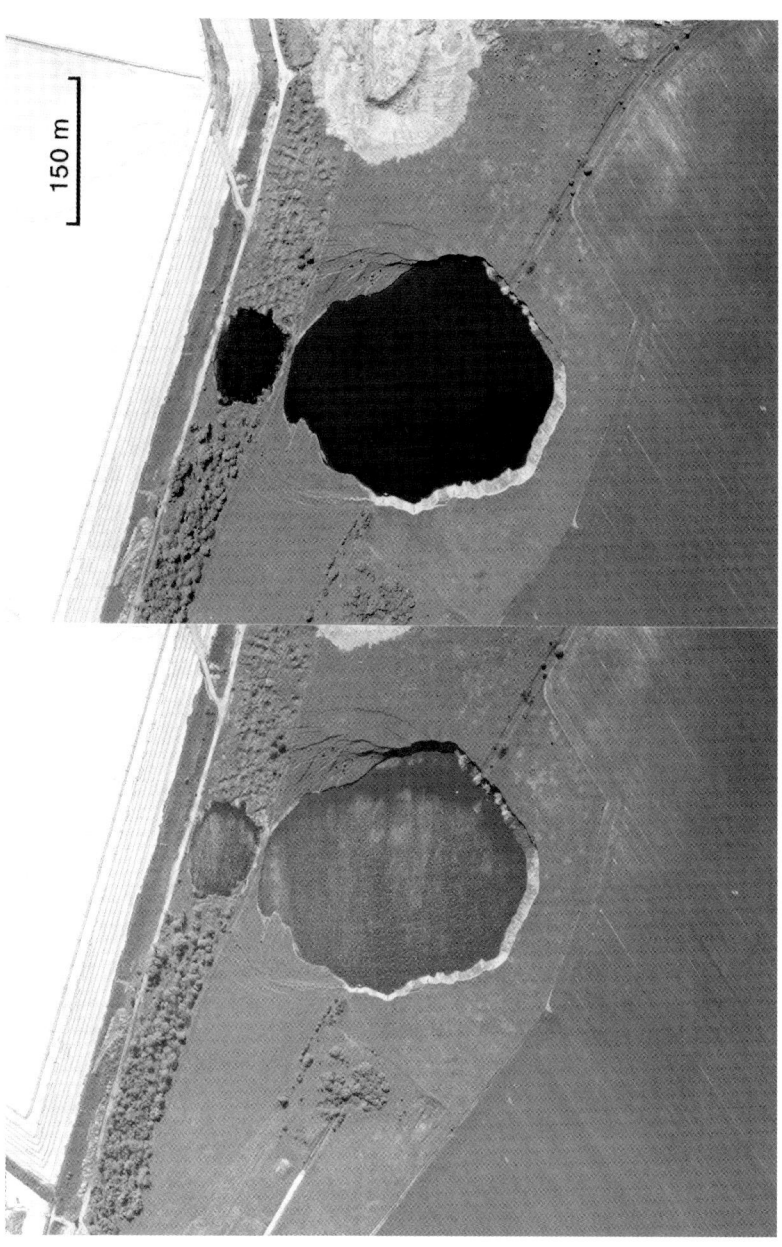

Fig. 4.12. Black-and-white aerial photograph (*stereopair*) can be interpreted to predict whether the waste heap will become unstable. Circular features at the edge of the collapse area (⌀ ca. 270 m) signal ongoing collapse in the direction of the heap (taken 08.05.94, by Berliner Spezialflug, Luftbild GmbH for BGR)

Vitality of Vegetation

Spectral range (type of data): *VIS (CIR AP)*, MS (scanner)

Plants are very good indicators of the presence of hazardous substances in air, soil, and ground water. The CIR aerial photograph taken in July 1990, shows terrain east of the Vorketzin waste disposal site in Brandenburg, Germany (Fig. 4.13). An example of environmental damage can be seen in a row of poplar trees (*arrow*), with impairment increasing from the lower to the upper part of the photograph. Contaminants from the neighboring waste disposal site are suspected as the cause of the impairment; however, ground checks are necessary to confirm these suspicions. Methods for assessing the vitality of vegetation using CIR aerial photographs are discussed in Section 6.3.2.3.

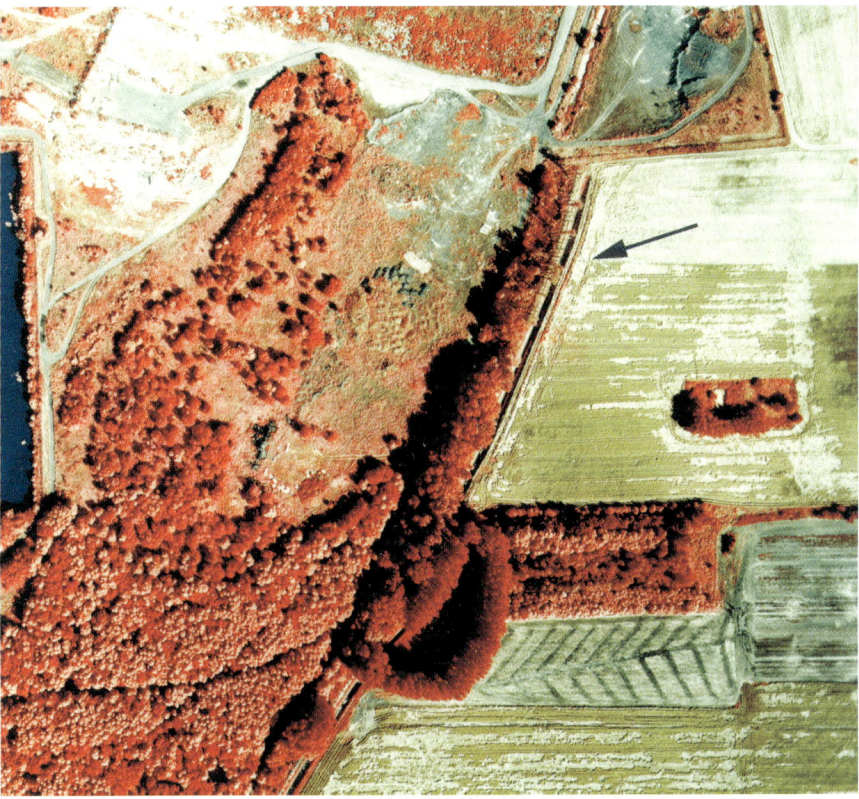

Fig. 4.13. CIR aerial photograph of the Vorketzin waste disposal site in Brandenburg, Germany. Increasing impairment of the vitality of a row of trees (*arrow*) with decreasing distance from a waste disposal site is indicated by the decrease in intensity of the red color. Original scale: 1 : 5 000 (photograph taken July 28, 1990, by Berliner Spezialflug, Luftbild GmbH; printed courtesy of Gesellschaft für Umwelt- und Wirtschaftsgeologie mbH, Berlin)

Assessment of Surface-Water Conditions

Spectral range (type of data): *VIS* (BW, *CIR AP*) MIR-II (Scanner), *MS (scanner)*

Water percolating through a landfill dissolves any soluble substances in the waste. If the layers immediately below the landfill are permeable, highly mineralized water will seep into the adjacent soil. This water will enter the groundwater and be transported along with the natural ground water flow. Water from the contaminated aquifer may enter surface water in the vicinity of a waste disposal site. In many cases, pollution of surface water is indicated by changes in the color of the water in the CIR photographs. The blue in the water body in the CIR aerial photograph of Fig. 4.14 indicates intensive turbidity (see Fig. 3.4). This turbidity was determined by a ground check to be caused by illegal discharge of liquid manure from a nearby farm.

Fig. 4.14. CIR aerial photograph of water-filled clay pits of a former brickyard north of Ketzin, Brandenburg, Germany. The blue indicates the discharge of liquid manure (water bodies are normally black in CIR photographs) (photograph taken July 28, 1990, by Berliner Spezialflug, Luftbild GmbH; printed courtesy of Gesellschaft für Umwelt- und Wirtschaftsgeologie mbH, Berlin)

Middle Ages Settlement

Spectral range (type of data): *VIS* (*BW, CIR, Color AP*) NIR-I (IR AP), MS (scanner), MIR-II (scanner)

The concentric rings visible in Fig. 4.15 are a relict of the wooden palisade of a Middle Ages settlement. The dark rings are due to elevated concentrations of humus in the soil, which probably originated from rotting wooden supports or a ditch system in the fortification. The distinct ring structure is easily visible on the aerial photo, because at the time the photograph was taken, the humus accumulation

Fig. 4.15. Color aerial photograph taken June 21, 1991, showing traces (concentric rings in the center of the photo) of an Middle Ages settlement north of Magdeburg, Sachsen-Anhalt, Germany (photo: F. Boeker and F. Kuehn, BGR)

promoted the growth of a winter crop (dark green). The inner dark ring shows patches of dark green (e.g., *arrow*). Accumulations of pottery fragments were found by ground checks there. As explained by archaeologists, wastes were dumped at special places in Middle Ages settlements. The dark green patch (*arrow*) may indicate one of these waste sites. This example clearly shows that man-made changes to the terrain can be detected in an aerial photo even after hundreds of years.

4.2.3
Subsurface Characteristics of a Waste Disposal Site

Remote-sensing data predating a landfill can be used to estimate the subsurface characteristics of the landfill. Under certain conditions, current remote-sensing data of the area around the landfill may be extrapolated to below the landfill. Archived aerial photographs are normally the best source of this information. Shallow geophysical methods could also be used to estimate the subsurface characteristics, but at greater expense and limited capability to characterize the layers below the landfill.

This archival material can be used to characterize the subsurface of an active waste disposal site using standard photogeologic interpretation methods. Some important characteristics, such as depth to the water table, near-surface rock permeability, and the location of faults and fractures, can sometimes be interpreted from the pre-landfill photographs.

Archival Photographs Showing Excavation at the Location of a Current Landfill

Spectral range (type of data): *VIS* (*BW*, CIR, Color *AP*) NIR-I (IR AP)

Numerous waste disposal sites are in natural or man-made holes in the ground, such as abandoned quarries, clay and gravel pits, sinkholes, and caves with openings at the surface. The sides and bottom of these "holes" may expose permeable layers that provide ready access for contaminated leachate from the waste.

An aerial photograph from 1991 (Fig. 4.16) depicts an abandoned waste disposal site south of Ludwigslust in Mecklenburg-Vorpommern, Germany. The steropair from 1953 (Fig. 4.17) shows that the site was formerly a gravel pit that has since been filled with waste. Leachate from the waste site has direct access to the uppermost aquifer exposed in the sides of the old pit without the benefit of any natural filtering through soil, which would have occurred if the waste site had been placed on the natural surface instead of in a man-made depression.

The depth of the abandoned pit (and thickness of the waste) was determined by interpretation of the 1953 stereopair, as well as the areal extent of the old workings and the quantity of waste (see also Eulenberg example in Section 6.2). If the archived photographs had not been available, the subsurface characteristics of the waste site could have been determined using shallow geophysical techniques, such as DC resistivity and shallow seismic methods, but at much greater expense and time.

Fig. 4.16. Aerial photograph (*stereopair*) taken July 7, 1991, showing a former waste disposal site (*arrow*) southeast of Ludwigslust in Mecklenburg–Vorpommern, Germany. The presently covered and partly overgrown waste site extends slightly above the terrain surface (printed courtesy of the Amt für Militärisches Geowesen in Euskirchen, Germany)

CHAPTER 4 · The Use of Remote Sensing in Waste Disposal Site Investigation

Fig. 4.17. Archive aerial photograph (*stereopair*) taken May 28, 1953, showing a former deep gravel pit (*arrow*) at the location of an abandoned waste disposal site southeast of Ludwigslust (cf. Fig. 4.16). Direct dumping of waste into the pit has provided direct pathways for contaminated leachate into the uppermost aquifers (source: uve GmbH Berlin)

4.2.4
The Search for New Waste Disposal Sites

Remote-sensing data, particularly when combined with more traditional geologic and geophysical tools, can be used during the planning and development of new disposal sites. Information derived from remote-sensing data should be used in the early stages of waste site development to provide a regional geologic context to the potential site and help focus detailed studies on possible problem areas. The remote-sensing data needs to be integrated with other geoscientific data at each successive step in the development process.

During the evaluation of site locations, remote-sensing technology can help identify and evaluate the following:

a   Existence of natural and artificial pathways that could facilitate seepage from future waste disposal sites, including
    - soil substrate with a high permeability and porosity,
    - channels or troughs (dorrs, grabens) filled with porous material that might funnel leachate from planned disposal sites,
    - faults and joint systems,
    - man-made drainage systems;
b   Impermeable rocks at the location of the planned waste disposal site;
c   Shallow ground water table at the planned site, springs and marshy areas;
d   Possible leachate migration pathways into protected areas;
e   Previous human activity that might increase permeability and lower filtering and sealing effects of the rocks underlying the planned waste disposal site:
    - former gravel pits, stone quarries, building pits, etc.,
    - subterranean installations, old wells and shafts, buried ditches and tunnels used for pipe laying, etc.

As discussed previously, remote-sensing data can be best utilized when evaluated by a remote-sensing expert. Interpretation skills are gained through repeated use and applications of data from various sensors. However, field checks must be part of the data collection and evaluation process to ensure the accuracy and, therefore, utility of the data.

Chapter 5

Verification of Remotely Sensed Data

Trude V.V. King · Roger N. Clark

5.1
Introduction

Ground or field checks are an important part of any remote sensing study and are necessary to provide an accurate and useful interpretive product. Field checking is necessary to confirm the validity of spectral, spatial, and morphological interpretations. In general, field checking should be done during all stages of any type of a remote sensing investigation. The methods and magnitude of work necessary to complete the field checking will be dependent on the type of remote sensing data to be verified and the scientific questions to be answered. Remotely sensed data provides an assessment of natural and anthropogenic features as they appear at the time of data acquisition, and possible changes between data acquisition and field checking must be considered.

Verifying historical aerial images can be a difficult, but very important task in a remote sensing study. It is especially important to attempt to authenticate potential links between complementary data sets which were collected at different times. The cultural and natural character of an area may have changed dramatically over time, requiring careful analysis of the data to establish commonality between the modern and historical images. The historical data may contain valuable information about physical or cultural conditions that have been obscured, changed or eliminated, but are crucial to ongoing studies. Establishing linkages between historical and current data is particularly vital when attempting to reconstruct events at a hazardous waste site where physical or environmental modifications have taken place.

5.2
Virtual Versus In-Situ Verification

There are two types of verification of remote sensing imagery information: virtual and in situ. Virtual verification can be done by examining the remote sensing data directly: there is sufficient spatial and/or spectral resolution to positively identify objects in the image. In situ verification requires visitation to the area of interest and direct sampling of the environment to verify the remotely sensed information. The following are examples of the application of the methods of verification.

Example 1. Red cars. Consider a hypothetical problem of needing to determine the number of red cars in a parking lot using remote sensing data. It is essential to separate cars from other vehicles, as well as distinguish the specific color. If

Landsat TM data were the only available data set, with 6 visible to near-IR bands and 30 m pixels, it would be impossible to resolve car shapes. In addition, a spectral unmixing algorithm to find pixels containing red signatures would be necessary. However, if an analysis was applied, the resulting classification image would not be positive identification of red cars. In addition, the field verification would be needed to confirm that the derived red areas actually contained red cars. In situ field checking would have to be done nearly simultaneously with the data acquisition because the distribution of cars in the parking lot might change. To verify the image information, red cars would have to be accurately located in the parking lot and on the image data to assess the accuracy of the classification. If the classification is shown to be accurate, the classification scheme could be extrapolated to other parking lots to give an indication of the number of red cars in those lots.

Cars could be distinguished from other vehicles on high resolution black and white imagery but the cars could not be determined with 100% accuracy. Subtle differences in the gray scale might allow red cars to be distinguished with some

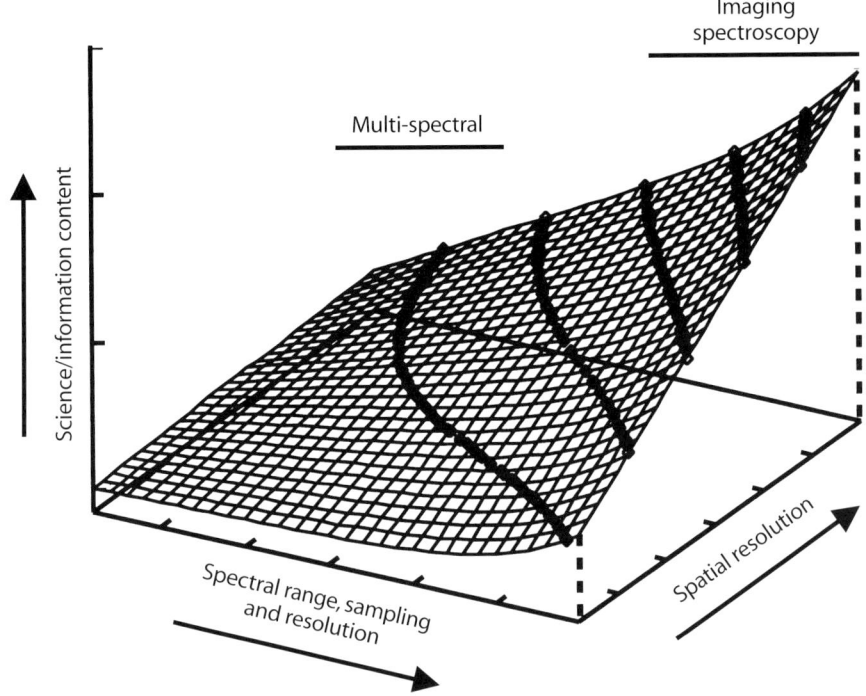

Fig. 5.1. The information content of remote sensing data is a function of spectral range, sampling and resolution and spatial resolution. Although narrow sampling intervals, high spectral resolution and a wide spectral range improve the information content in remotely sensed data, similarly so does high spatial resolution. For optimum science information, ideally both the spectral and spatial parameters would be maximized

degree of reliability. In-situ verification would need to be done nearly simultaneously with the data acquisition to ensure accuracy in the identification of red cars. However, color imagery with sufficient spatial resolution to distinguish cars from trucks, and sufficient spectral resolution to discriminate the color red, would allow all red cars in the parking lot be identified with 100% certainty without any in situ field checking. Simple examination of the imagery would allow red cars to be identified and counted much the same as would be possible by being in the parking lot at the time the data were acquired. Such positive identification gained directly from the remotely sensed data is called virtual verification.

Note that there is a distinction between "identification" and "classification" of results from the analysis of the remote sensing imagery. While some objects can be identified directly from the imagery, others can only be inferred to some level of confidence (classification) that will need in situ field checking. Information gained from the analysis of remotely sensed data increases with increasing spatial resolution and/or spectral resolution and, in general, is maximized when both high spectral and high spatial resolution are used together (Fig. 5.1). Although, in some instances only high spatial or spectral resolution is needed. For example, identifying cars requires high spatial resolution, but only poor spectral resolution. Black and white imagery would suffice to identify cars; a simple color photo could be used to determine its color.

Example 2. Minerals in Soils. If the problem is to identify the presence of a specific mineral, for example calcite, in a large areal exposure of soil, the following should be considered. The use of black and white imagery would provide little chance of locating calcite. Extensive field sampling, including returning samples to the laboratory for analysis, such as X-ray diffraction, would be needed to derive a correlation of ground albedo (image gray level) with calcite occurrence. However, such a classification is not a positive identification as the possibility exists that materials other than calcite could produce a similar response.

Using imaging spectroscopy data, with suitable spectral resolution, it is possible to identify specific minerals in soils, such as calcite, based on the wavelength position and shape of characteristic absorption features (see 6.7 for more detailed discussions). The detection of unique calcite spectral absorption features allows the positive identification of the mineral and the capability to map its distribution, based on the limits of the spatial resolution of the instrument. In this case, there is no need for in-situ field checking because the spectra are of sufficient resolution to be certain of their identification. The derived calcite maps can be verified by examining spectra from the imaging spectrometer data, a form of virtual field verification.

Increasing the spatial resolution of the instrument would not increase the likelihood of mapping the presence of calcite in the soils, unless the spatial resolution approached that of the grain size of the individual calcite grains. If the spatial resolution approached that of a microscope, individual crystal morphologies could be resolved and the areal distribution of a specific mineral could potentially be mapped using this criteria. Consequently, spectral resolution is more important than spatial resolution for identifying specific mineral composition.

5.3
Verification of Vegetation Data

Verification of vegetation spectra is challenging whether the in-situ or virtual approach is used. During the verification process, the spectral signatures of specific plants or plant communities are correlated with the spectral signatures in the remotely sensed data. The absorption features of interest result from the internal cellular structure and chlorophyll, ligand, and water content of the vegetation. Thus, spectra used in comparisons should be from plants having analogous growth cycles and environmental conditions as the remotely sensed plants. In-situ verification should be completed as soon as possible after remote data acquisition, minimizing the spectrally detectable natural chemical and climatic responses resulting from changes in environmental conditions and natural growth cycle. See Kronberg (1985), and Kuehn et al. in Section 6.3, and King et al. in Section 6.7 for further discussions.

The problem of viewing aspect adds to the complexity of verification of remotely sensed vegetation information. Remotely sensed data is collected looking downward, similar to what an individual can see from an airplane. The overhead viewing position influences the proportions of leafy green material, bark, and stems seen in individual vegetation species. Consequently, in forested areas the remotely sensed data measures the canopy characteristics, rather than characteristics of individual trees and plants, some of which may grow beneath the canopy.

Logistically, verifying the composition or structure of the vegetation canopy is difficult because of cost or access issues. In-situ verification measurements, helicopter based for example, are desirable, but in most instance they are extremely expensive and often prohibited by access issues. Consequently, verification must be done by comparing weighted spectra of individual components- varying proportions of leaf, stems and bark- to the remote canopy observations.

High spatial resolution data, color aerial photography for example, can sometimes resolve the crown shapes of trees which can be used to identify specific species. Despain (1990) used crown shapes in color aerial photographs to map trees in Yellowstone National Park, but bushes and grasses were too small to resolve the structural detail related to species. More recently, Kokaly et. al (1998) used *AVIRIS* data to define tree, shrub, and grass species based on their spectral signatures, although crown shapes were not resolved.

Regardless of the type of remote sensing data, virtual or in-situ verification will improve the accuracy and usefulness of the final data product.

Chapter 6
Case Studies

6.1
Introduction

The following case studies are intended to illustrate possible uses of remote sensing data for the investigation of waste disposal sites. The selected case studies represent a broad thematic spectrum. The approaches range from conventional aerial photograph interpretation to modern hyperspectral techniques. This English-language version of the German edition has been expanded with representative contributions from the United States and Canada. The comparison of case studies from North America and Germany shows how the difference in conditions lead to different methodological approaches.

The German case studies deal exclusively with waste disposal sites. Except for the Eulenberg waste disposal site near Arnstadt, which has a rather unusual prior history, the sites generally display typical problems of site investigations. These problems include incomplete data sets, nonoptimal data acquisition times, poor quality of some older aerial photographs, and complicated landscape features. The following aspects are the main interest when hazardous waste contaminated sites are investigated:

- existence, state, and extent of a geologic barrier at the base of the landfill,
- possible migration paths for contaminated water from the landfill,
- internal structure and peculiarities of the landfill.

The case studies outline the advantages and limitations of the use of remote sensing data to address disposal site problems. The following examples of applications have been chosen:

- Combination of an analysis of the site history, photogeologic interpretation, and geophysical investigations (Eulenberg waste disposal site near Arnstadt in Thuringia) by Friedrich Kuehn,
- Evaluation of a landfill site using remote sensing data from different sensors, and acquisition dates (Schoeneiche landfill near Mittenwalde in Brandenburg) by Friedrich Kuehn, Bernhard Hoerig, and Dietmar Schmidt.

Summarized or simplified results and maps are presented to illustrate the respective methods and work routines.

The North American case studies also deal with regional and local environmental characteristics and the application of remote sensing techniques. Similar to the

German projects, the location and characterization of contamination sources, investigation of pollution pathways and of pollutant accumulations in the vicinity of the sources, and risk assessment for the respective regions are the major topics of these sections. Methodology and results are given for the following applications:

- Combination of aerial photo interpretation and thermal infrared remote sensing to locate hidden landfills with a high risk potential (Solid Waste Storage Area 4 at Oak Ridge National Laboratory, Tennessee) by John M. Irvine et al.
- Application of satellite and airborne data to characterize and monitor mining and tailing sites (mining-related environmental impact in Sudbury, Ontario) by Vernon Singhroy.
- Application of satellite and hyperspectral airborne data to characterize mine wastes (mining areas at Cripple Creek, Colorado, and Goldfield, Nevada) by Douglas C. Peters and Phoebe L. Hauff.
- Application of hyperspectral remote sensing to investigate a mining site and its areal impact (former gold mining area of Summitville, Colorado) by Trude V.V. King, Roger N. Clark, and Gregg A. Swayze.

Information on the affiliations of the authors and their addresses is given in the list of contributors at the beginning of this book.

6.2
Archival Aerial Photographs Used to Evaluate the Subsurface of Waste Disposal Sites (Arnstadt, Germany)

Friedrich Kuehn

6.2.1
Introduction and Problem Description

Unusually complicated conditions below the Eulenberg landfill near Arnstadt make this site of particular interest for testing for methodological investigations. Using this site as an example, the following discussion will demonstrate the advantages of a combined application of geoscientific methods (see also Kuehn et al. 1994).

The Eulenberg waste disposal site was in use from the early 1960s until 1979. The site is on a north-facing slope of 15–20°. The landfill was started in the pit of an apparently unfinished underground bunker from the days of World War II. The exact location of the original excavation and the layout of the subsurface bunker system were unknown at the beginning of the study.

Today, the landfill extends beyond the excavation site and only part of this expansion lies on relatively impermeable clay strata (Keuper). Most of the site overlies marly limestone (Upper Muschelkalk), which has a pronounced jointing with a NW–SE orientation and is an aquifer.

In addition, the site is crossed by the WNW-trending Eichberg-Gotha-Saalfeld fault zone, which is a deep-seated Saxonian fault system in the southern Thuringian basin (Wunderlich et al. 1991, 1992). It also lies within the protection zone of the Schoenbrunn waterworks.

The landfill contains about 800 000 m³ of unsorted domestic and industrial wastes 10 m to 25 m deep in an area of more than 6 ha. The wastes consist of building rubble, ash from home furnaces, slaughterhouse wastes, tanning residues, galvanizing and grinding sludge, sewage, cyanide-containing sludge, and phenol-containing waste from asphalt production.

A lack of knowledge about the thickness, extent, and condition of the impermeable strata below the landfill made it necessary to conduct a comprehensive investigation of the geologic structures and the hydrogeologic situation, including the hydraulic pathways for potential pollutants. The combination of the natural geological situation and the possibility that the layers at the base of the landfill were no longer intact made the planned investigation particularly complicated.

The current conditions at the site had to be taken into consideration during the investigation of the Eulenberg waste disposal site. Large parts of the area have been built on since landfill operations were terminated (see Fig. 6.1). Profiles for shallow geophysical investigations could not be laid out straight across the landfill, but had to follow the morphology of the landfill.

In addition, the rock layers at the base of the landfill are covered by a thick, heterogeneous layer of building rubble and excavated soil and rock, which appear in reflection-seismic sections as a "disturbance". The condition of the natural lay-

Fig. 6.1. Color Infrared (CIR) aerial photograph of the Eulenberg waste disposal site near Arnstadt, Thuringia, taken August 7, 1991, showing the position of the seismic reflection profile (*Rx*) of Fig. 6.2; the profile crosses the landfill, which is approximately centered in the photograph (photo: Hansa Luftbild Ltd for the County of Arnstadt)

ers below the landfill and the exact location of the remains of a subsurface building (unfinished production bunker from WW II) were extremely difficult to reconstruct solely on the basis of geophysical data. Consequently, interpretation of aerial photographs was also necessary.

6.2.2
Geophysical Investigations

Selected examples of the geophysical investigations (seismics, gravimetry, geoelectrics and magnetics) conducted at the site are briefly being outlined in this section (cf. Kuehn et al. 1994). The aim of the seismic reflection measurements was to investigate the base of the landfill and the geologic structure beneath it. The seismic reflection data reveal the structural and lithologic configuration, starting at depth of about 4 m below the landfill surface down to a depth of more than 150 m. Figure 6.2 shows a cross section through the western part of the waste disposal site (see Fig. 6.1). The reflections of the generally northeast-dipping layer of Muschelkalk rock are relatively well defined. The reflections below the Middle Muschelkalk correlate with Lower Muschelkalk and Bunter.

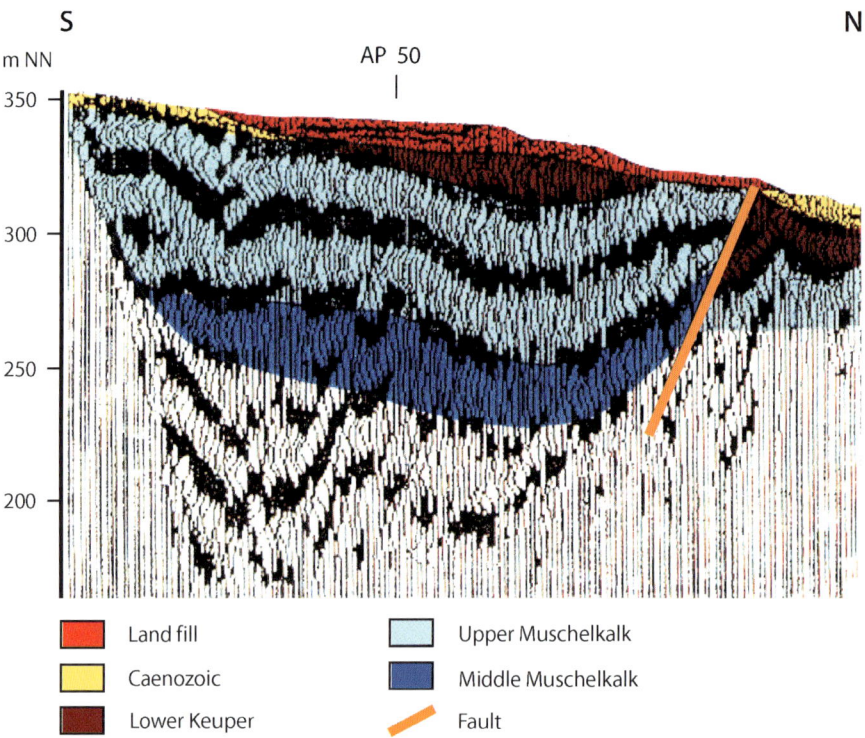

Fig. 6.2. Seismic reflection section through the Eulenberg landfill; carried out by Geophysik GGD for BGR (Meyer 1993). For the location of the profile, see Fig. 6.1

Offset related to the Hercynian Eichenberg-Gotha-Saalfelder fault zone is detectable in the northern part of the seismic section. However, no disturbance of Keuper and Muschelkalk layers that could be related to the construction of an underground bunker is observable. The distances between the seismic profiles and the spatial resolution of the seismic data were not suitable for detecting such relatively small-scale features.

Because the seismic investigations could not locate man-made disruptions of the sealing layers below the landfill, it was attempted to obtain this information from gravity field measurements. A potential density contrast of ca. 800 kg m^{-3} was expected due to difference in the densities of the Keuper and Muschelkalk rocks ($\rho \approx 2300$ and 2500 kg m^{-3}, respectively) and of the waste material ($\rho \approx 1600$ kg m^{-3}). Thus, a density minimum was expected at the location of the underground bunker, which is filled with waste. The Bouguer-anomaly map (Fig. 6.3) shows a large area of low values. The significant anomaly (dark blue) in the right center of the gravity map can be explained by a large thickness of low

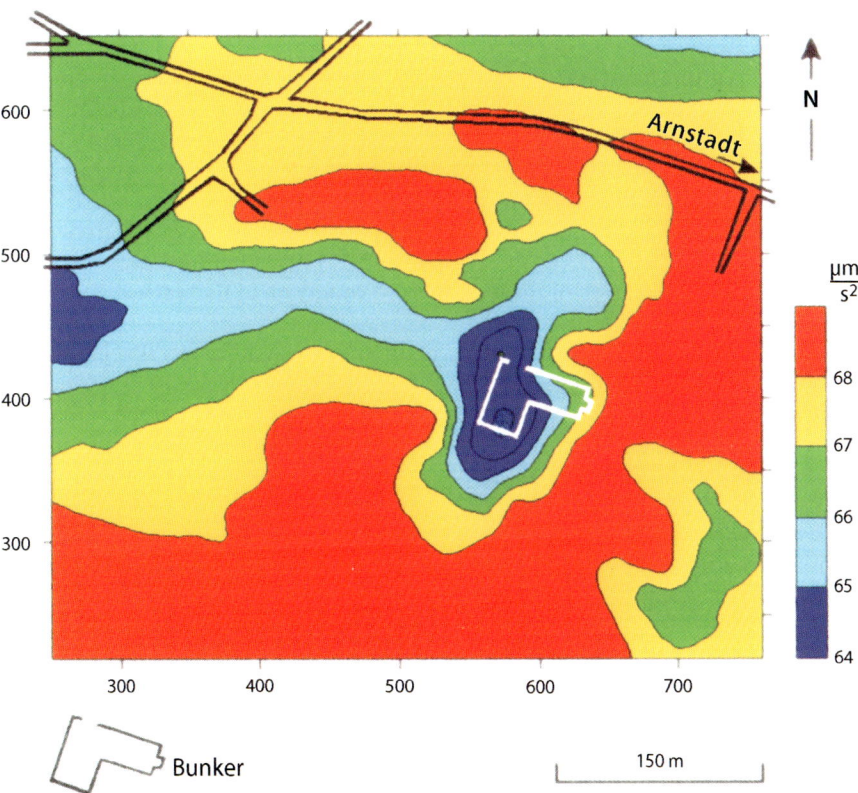

Fig. 6.3. Gravimetry (Bouguer anomaly) in μm s^{-2} with the outlines of a bunker that was later detected in aerial photographs (compiled by Geophysik GGD, Leipzig, on behalf of the BGR; after Schulze et al. 1992)

density waste as would occur by dumping the waste in a deep pit. Underground construction and disruptions of the sealing layers at the base may be expected in this part of the landfill, but they could not be detected by gravimetry.

Results of the magnetic and geoelectric investigations, which led to similar results, will not be discussed here (cf. Kuehn et al. 1994). The use of several geophysical evaluation methods yielded a broad spectrum of physical properties of the rock below the landfill.

For detailed information about the base of the landfill, geophysical methods have their limits. Smaller distances between geophysical profiles or grid points would increase the resolution, but would not sufficiently clarify small disturbances of the layers below the waste. Moreover, more extensive geophysical studies would increase the cost of the investigation. The results presented in the following section show that the interpretation of aerial photographs can provide a more detailed interpretation of the site and could have contributed to a better selection of the methods employed.

6.2.3
Interpretation of Aerial Photographs

Because disruption of the rock layers at the base of the landfill occurred in conjunction with the construction of an underground bunker during World War II, aerial photographs from those days were examined. In 1944 and 1945, the Arnstadt/Eulenberg site was the target of American reconnaissance flights several times. Aerial photographs from those flights and target information sheets, made by the reconnaissance pilots for the bomber pilots, were available from US archives. A series of aerial photographs from German topographic mapping flights was available for the reconstruction of the location and condition of the waste disposal site from 1945 to the present.

The significance of the archival aerial photographs lies in the fact that large parts of the area have undergone drastic changes, and ground checks are now possible only in some cases. Thus, the archival aerial photographs are evidence of conditions in the past, the truth of which depends on the photographic quality of the aerial photographs, the recognizability of features in photographs made on different flights, and the experience of the interpreter.

Even if part of the information is uncertain, the photographs may yield important information on the areas where material was removed from the rock layers underlying the waste disposal site, on the probable extent of the bunker, as well as on the location of entrances to finished parts of the bunker. In addition, knowledge concerning the distribution of steel-reinforced concrete underground facilitates the interpretation of the geophysical data. The aerial photographs provided an invaluable view of the area and parts of the underground construction before waste disposal was begun in the 1960s and 70s (Fig. 6.4). Figure 6.5 shows a sketch drawn from stereoscopic evaluation of the aerial photographs.

According to information derived from aerial reconnaissance photographs, and from other sources, the bunker consisted of partly finished rooms for underground laboratories and production. Precision instruments, control systems, and secret weapons systems are thought to have been made there, and up to 5 000 workers employed.

In 1945, the most obvious site characteristic is a deep pit containing the obviously unfinished bunker. The building had an L-shape and had outer dimensions of ca. 55 m × 70 m. It is not recognizable whether the long eastern wing visible in the photograph is the roof of the bunker or a foundation.

Comparison of the photographs from 1944 and 1945 suggests that construction had been suspended or was being continued very slowly. The high water level in the pit suggests that construction had been suspended. It can be seen in the stereoscopic view that the water in the small pit northeast of the main pit (*arrow* in Fig. 6.4d) is at a higher level than in the main pit. It is, therefore, assumed that as the excavation of the small pit advanced to the northeast it entered the east–west-trending, water-bearing fault system recently determined by seismic measurements. Flow of water from this fault system into the pit probably led to discontinuation of construction. Under present site conditions, contaminated water from the landfill enters the fault system via the pit and, thus, into the area around the waste disposal site. This possibility led to a hydrogeologic analysis, which had not been completed when this book went into print. This example shows that the engineering and geologic factors have to be assessed together to produce an adequate environmental–geological assessment.

Traces of further wings of the underground bunker, which are possibly still accessible today, can be detected on the west side of the pit where a large opening is visible (width ca. 6 m, height ca. 3 m). This expansion was likely a planned link between an already functioning underground installation and the construction in the pit. The entrance is at water level but remains dry, due to a clearly recognizable artificial barrier. Indications of other tunnel openings are discernible in the vicinity of a trail around the pit.

When evaluating openings visible on the photographs, the possibility has to be considered that fake entrances could have been created to make aerial reconnaissance more difficult. In this case, information about alleged laboratories and production rooms, the barrier in front of the opening, signs of extensive earthwork around the pit, and an already established infrastructure, support the possible existence of large hollowed-out spaces west and south of the pit.

In the aerial photographs from 1961, in which the appearance of the site is similar to the way it looked in 1945 with the exception of missing buildings, the former tunnel openings have clearly been destroyed. As mentioned above, a zone of extensive earthwork is recognizable around the pit. It can also be seen that excavated material was dumped in this zone south of the pit in an area that had been excavated below the original surface. Thus, it can be concluded that this zone may have been the site of a construction pit for another, finished part of the bunker. Further evidence for this conclusion is a camouflaged building not far to the south of this zone. Experience with similar facilities has shown that underground bunkers generally have emergency side exits. The geomagnetic data revealed high values for the total intensities in the area between the pit and the camouflaged building, which only to some extent correlate with the contours of the waste heap and may reflect the presence of a buried steel-reinforced concrete tunnel between the main bunker and the assumed emergency side exit.

Traces of extensive disturbance of the ground can also be detected to the west of the pit. Whether this disturbance is due to the backfilling of a pit cannot be

Fig. 6.4 a,b. Photographic history of the Eulenberg waste disposal site at Arnstadt, Thuringia, with aerial photographs from 1981 (**a**), 1971 (**b**) (photographs from 1971 and 1981: KAZ Bildmess GmbH Leipzig)

Fig. 6.4 c,d. Eulenberg waste disposal site 1961 (**c**), and 1945 (**d**, *stereopair rotated 50°*) (archives research and acquisition of aerial photographs from 1945: Luftbilddatenbank Würzburg; photograph from 1961: KAZ Bildmess GmbH Leipzig)

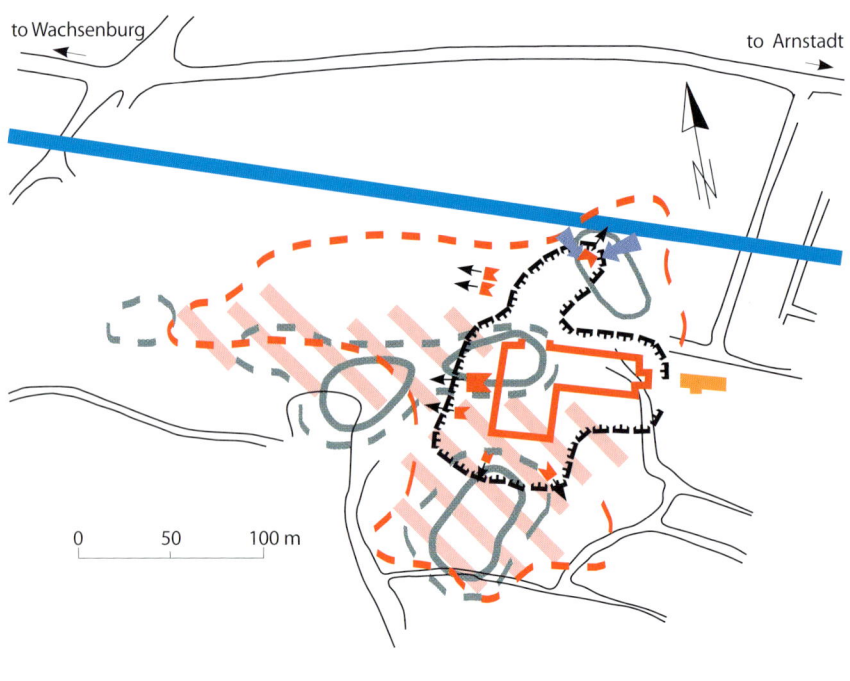

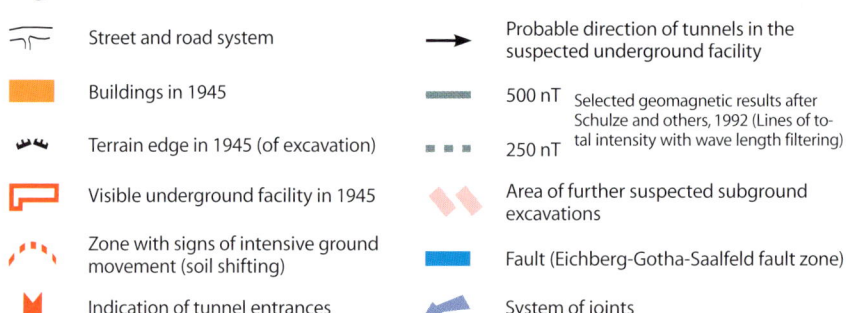

Fig. 6.5. Simplified map of the interpretation of archival aerial photographs and geophysical data from the Eulenberg waste disposal site at Arnstadt in Thuringia (from Kuehn 1992)

determined from the aerial photographs alone. This area, which was not anomalous in the seismic investigations, became an area of interest when the gravimetric results showed a distinct gravity minimum in this area (see Fig. 6.3). The possibility that further excavation pits may have existed to the west and south of the pit visible in the old aerial photographs must be taken into consideration because they can act as potential seepage water pathways from the landfill to the surrounding area.

6.2.4
Summary

Although the example of the Eulenberg waste disposal site at Arnstadt surely is not a typical case for municipal waste dumps, it has been demonstrated how to characterize a complex site situation with the combined application of aerial photographs and geophysical exploration techniques.

The challenge was to reconstruct the previous site conditions, which were quite different from the present situation. In addition, the geologic situation at the base of the landfill beneath a layer of waste as thick as 25 m had to be investigated. In this case study, anthropogenic alterations and natural geologic conditions at the site determine the potential environmental threat from this site. Considerable information was obtained by combining geophysical exploration techniques with interpretation of aerial photographs. For the Arnstadt/Eulenberg waste disposal site, information both on underground features beneath the present-day landfill and on its geologic base were obtained. On the basis of this knowledge, possible migration pathways for pollutants from the landfill to the surrounding area could be projected.

The experience gained from this study suggests that site characterization should begin with interpretation of archival aerial photographs. This would provide basic information about the character of the site and the ground below the present landfill, and lead to effective application of ground-based exploration techniques (geophysics, drilling, etc.).

6.3
Airborne Remote Sensing to Characterize Waste Disposal Sites (Schoeneiche, Germany)

Friedrich Kuehn · Bernhard Hoerig · Dietmar Schmidt

6.3.1
Introduction and Problem Description

Two large landfills, the Schoeneiche and Schoeneicher Plan, are located southeast of Mittenwalde near Berlin and are separated by a 200- to 300-m-wide strip of land (Figs. 3.1 and 6.6). The abandoned water-filled clay pits of the Schoeneicher Plan site were used by Berlin for waste disposal from about 1920 to the mid-ninties. The Schoeneiche site has been used since 1977 by West Berlin (since 1991 by reunified Berlin) for disposal of municipal waste, building rubble, and excavated soil. A clay seal was not installed. The landfill is partly covered by clay or synthetic foam (Piasol).

Both waste disposal sites lie in a wide glacial valley of Weichselian (Pleistocene) age cut into the basal till of the Teltow plateau. Erosional remnants of Weichselian and older Quaternary till form low relief features around the site. Humic sand of varying grain sizes, limnic silty clay, limnic marl, peat, and paleoculture material are found above the unconfined aquifer (Ahrens 1992).

Both landfills are located on level ground 0.5–1.5 m above the water table. The uppermost, unconfined aquifer consists of sand and is in hydraulic contact with a

Fig. 6.6. Oblique aerial photograph taken on May 11, 1993, looking westward across the waste incineration facilities at Gallun (*foreground*) and the northern part of the Schoeneiche landfill (*middle ground*) to the Schoeneicher Plan landfill in the background (photo: F. Boeker and F. Kuehn, BGR)

number of streams. Runoff from both sites passes primarily through numerous ditches that flow into the Notte Canal, north of the sites. The general geologic setting of the study area is shown in Fig. 6.7.

The lack of a seal (an impermeable layer, such as clay) at the base of the landfills allows water to enter the aquifer and then into the surface drainage system. Ahrens (1992) found contaminants in water samples from ditches north of the sites, although the origin has not been identified because there are several possible sources in the immediate vicinity. The operator of the Schoeneiche site is currently attempting to install a basal seal and a system to collect seepage water from the landfill.

The Schoeneiche and Schoeneicher Plan landfills were selected for study because the geologic setting is quite different from that of the Eulenberg landfill discussed in the previous section. The primary goal of this study was to investigate

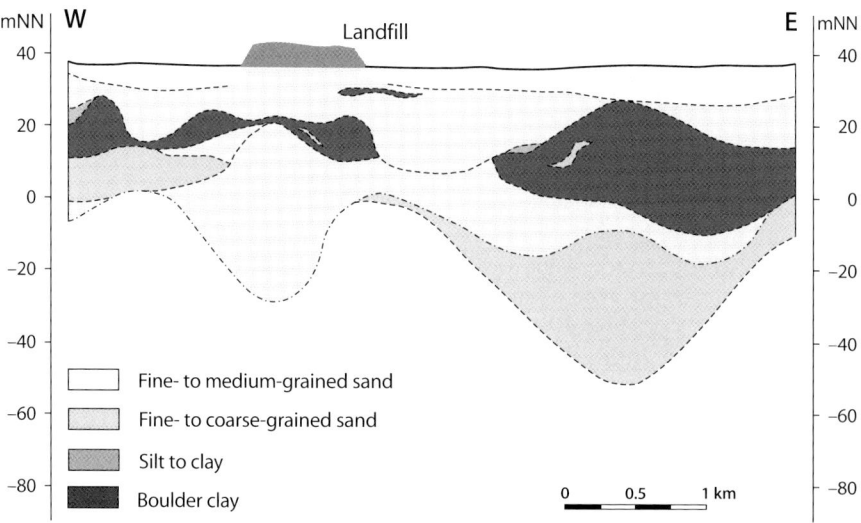

Fig. 6.7. Cross-section showing the geologic setting of the Schoeneiche and Schoeneicher Plan landfills (simplified after Parsiegla 2000)

the use and limitations of remote sensing methods in this geographic and geologic setting.

Archival aerial photographs, recent aerial CIR photographs, and *CASI* scanner and thermal scanner data, were evaluated. Groundchecks and data from conventional geologic and geophysical field investigations were included in the evaluation. This combination of investigative techniques proved to be a very effective, integrated approach for site characterization. Because of the close proximity of the Schoeneiche and Schoeneicher Plan landfills, both sites were included in this study. The goals of the investigations included

- description of soil in the uppermost layers at the sites,
- determination of the soil conditions in the vicinity of the sites,
- delineation of possible soil or ground water contamination outside the perimeters of the sites,
- determination of possible migration pathways for leachate outward from the sites,
- characterization of the waste, as far as possible.

During the study, it became apparent that it would be more effective to divide the area into smaller sections for data collection, analysis and interpretation. This facilitated identification of specific problems and appropriate strategies for data acquisition, interpretation of aerial photographs, image processing, and ground checks could be developed to address the site-specific problems at the landfill itself and at its surroundings. The results for the smaller sections were integrated for an assessment of the entire study area.

6.3.2
Interpretation of Aerial Photographs and Scanner Images

6.3.2.1
Examination of the Waste Disposal Site

Characterization of buried waste material using remote sensing techniques is difficult, and it is best to combine different remote sensing methods to relate surface expression to subsurface properties (e.g., variations in thermal patterns may indicate changes in the type of subsurface waste). If detailed records are lacking, approaches that include archival aerial photographs and maps from different times are required to determine the type and amount of material deposited in a landfill.

A recent CIR photograph of the southeastern part of the Schoeneiche waste disposal site is shown in Fig. 6.8. Stereoscopic analysis revealed areas of recent dumping, areas of surface water, and other features. Comparison of this photo-

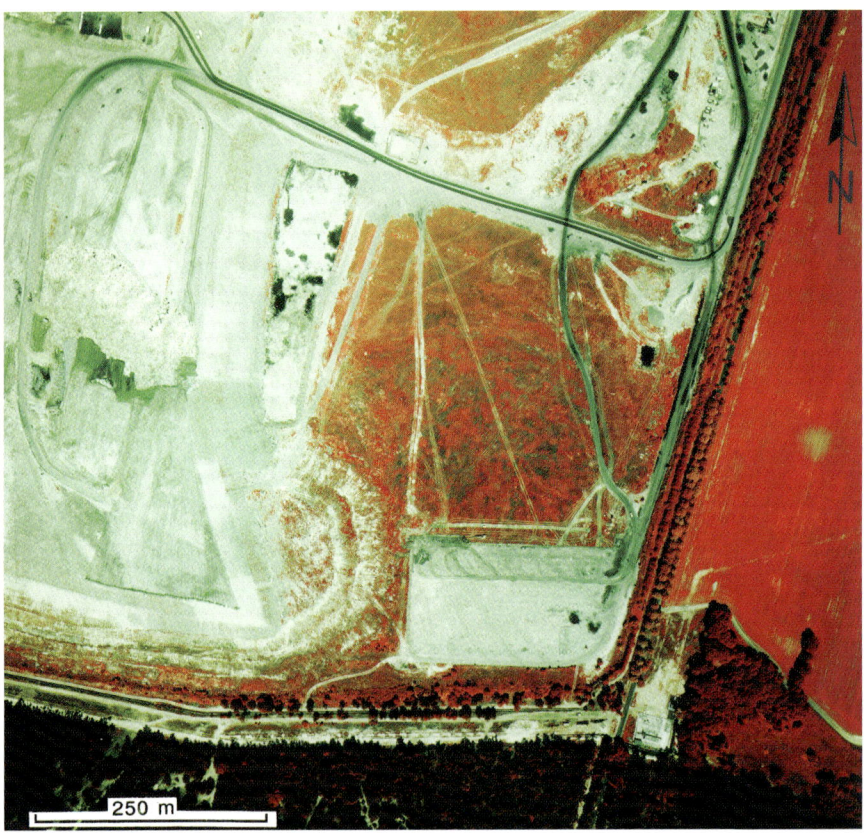

Fig. 6.8. Portion of a CIR aerial photograph taken on July 2, 1993, showing the southeastern part of the Schoeneiche waste disposal site, scale 1:5000 (aerial photograph: WIB GmbH for BGR)

graph with ones from previous flights provided additional information on the nature, type, and distribution of buried waste (cf. Section 4.2.1).

The Schoeneiche and Schoeneicher Plan waste disposal sites were scanned with the *AGEMA 900* thermographic system on May 11, 1993, at noon and on May 13, 1993, from shortly before until shortly after sunrise. These flights were designed to investigate the utility and limitations of thermal imaging. The flight altitude of both flights was 1 060 m above ground level (FOV = 20°, IFOV = 1.5 mrad) and there was no rain or significant change in the weather between the flights. Evaluation of data from these flights yielded the relative radiation temperature information in Table 6.1.

Previous investigations of waste disposal sites have shown that thermal states are related to the time of day, time of year, and the weather conditions prior to data collection; the useability and reproducibility of aerial photographs and thermal images are also related to these parameters. Qualifiers, such as warm or cold can be used only together with a specific time of day because of the diurnal temperature variations differ for different materials (Fig. 6.9). Consequently, generalized statements regarding the thermal properties of materials in the Schoeneiche landfill (e.g., Ehrenberg 1991) are of questionable value.

To maximize the usefulness of thermal remote sensing data, surface temperatures of rocks, soils, vegetation, and waste should be collected at the same time the remote data is acquired. If required to clarify specific questions, samples can be collected and their emissivities measured in the laboratory.

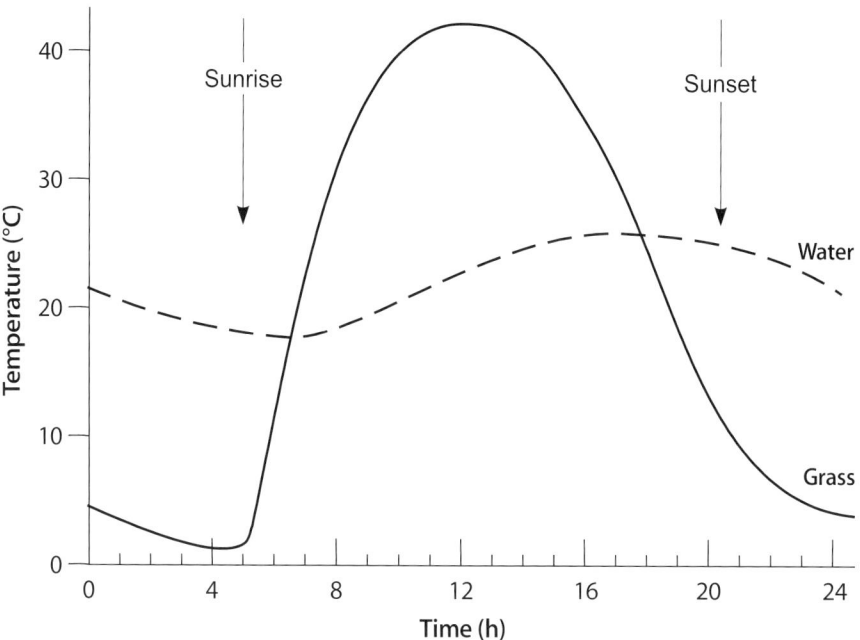

Fig. 6.9. Surface temperatures of water and grass during a diurnal cycle (after Lowe 1969)

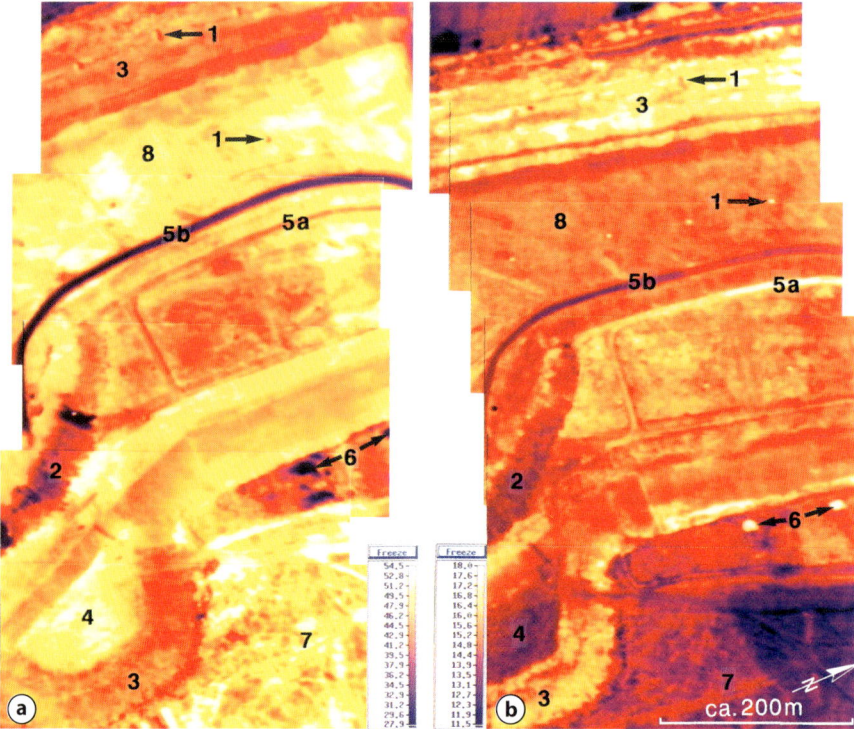

Fig. 6.10. Two thermal images of the southern part of the Schoeneiche waste disposal site. **b** nighttime image taken at 05:15 CET on May 13, 1993. **a** daytime image made at 13:00 CET on May 11, 1993. No precipitation or change in the general weather conditions occurred between the two flights (images: F. Boeker and F. Kuehn, BGR). The figures refer to numbered features discussed in the text (*pos. 1–8*). The temperature scales are in degrees Celsius

Figure 6.10 shows thermal images from nighttime and daytime flights over part of the Schoeneiche waste disposal site. Table 6.1 summarizes the relative temperatures of different materials at the surface of the site observed in this data. As illustrated in Fig. 6.9, the relative temperatures of certain materials at night may be reversed during the day. Comparison of the two thermal images in Fig. 6.10 illustrates diurnal variations, particularly of the surface water, vegetation, clay cover, gas ducts, and roads. The temperature of sloping ground depends on the aspect and the angle of slope. The following numbered features can be distinguished in Fig. 6.10 on the basis of relative temperatures:

1. **Gas emission ducts:** In the northern part of the Schoeneiche landfill, a network of ducts is used for gas collection and dispersal. Where the gas ducts emerge, they are distinguishable as "hot spots" at night, when the site surface cools off relatively rapidly. In the daytime, solar radiation heats up the surface and the ducts appear colder than the surrounding surface. These gas ducts are cur-

Table 6.1. Comparison of the temperatures of typical surfaces of the Schoeneiche landfill in each thermal image relative to the ambient temperature. Nighttime flight made on May 11, 1993, and daytime flight on May 13, 1993

Thermally conspicuous area	Nighttime flight	Daytime flight
Slopes and embankments with sunny exposure[a]	warm	warm
Ditches, depressions	cold	cold
Trees and bushes	warm	cold
Water	warm	cold
Moist soil	cold	cold
Synthetic foam cover (Piasol)	cold	warm
Fresh waste, uncovered	cold	cold
Old household waste, uncovered	ambient	ambient
Household waste, covered	cold	warm
Asbestos waste, covered	cold	cold
Household waste under 50-cm loam cover	cold	warm
Light sandy soil	warm/cold	warm
Loamy soil with sparse vegetation (grass)	cold	warm
Dark humic soil	cold	warm
Roofs[a]	warm	warm
Roads, unpaved	warm/cold	cold
Roads, asphalt	warm	warm
Gas emission ducts	warm	cold

[a] Thermal properties are different for light and dark colored roofing materials.

rently the only source of information on temperature anomalies and the related decay processes within the landfill (see also Fig. 6.12).
2. **Fresh waste:** Recently deposited waste appears colder than the surrounding areas in both nighttime and daytime images. These observations are not in agreement with the generalization by Ehrenberg (1991) that recent wastes are warmer than older waste. In each case, both the composition of the waste, and the thermal properties of the surrounding materials are responsible for the relative temperatures.
3. **Sloping areas:** Sloping areas facing the sun heat up more than horizontal or shadowed areas and, thus, appear in both daytime and nighttime images as relatively warm, depending on the waste material (e.g., west-facing slope on the far left).
4. **Piasol foam cover:** The light-colored surface of the synthetic foam (Piasol) reflects a significant proportion of incident radiation, which results in low radiation absorption and heating. These surfaces appear "cold" in nighttime images and "warm" in daytime images.

5. **Roads:** The thermal characteristics of roads vary considerably in the images. In general, roads act as thermal bridges because of soil compaction, which enhances heat transfer from the interior of the landfill to the surface. Thus, the idea that roads generally appear "warm" is only true for areas with a thin cover over the waste and only in nighttime images (Fig. 6.10b, 5a). In areas with a thicker cover, roads are seen as "cold" in both daytime and nighttime images (Fig. 6.10, 5b).
6. **Surface water:** Most open water bodies appear warmer than their surroundings in nighttime images and colder in daytime images, except for very polluted or eutrophied water.
7. **Household waste under a 30–50 cm thick cover of clay with patches of grass:** The clay cover is relatively impervious and poorly drained; consequently, the temperature relations in the thermal image are primarily influenced by the moisture content. The images, taken in the middle of May, show that the clay layer has high moisture. A ground check must be made to determine whether this is because rain water is prevented from entering the waste or is due to accumulations of leachate.
8. **"Plateau" area:** Most of the landfill has a more or less horizontal, slightly undulating surface. These areas appeared relatively warm but inhomogeneous in daytime images, and more uniform in the nighttime images. Prominent thermal features were not observed.

These examples illustrate that under normal conditions, it may be quite difficult to derive information about heat sources within the waste or on particular properties of the waste by surface examination. The nighttime thermal image in

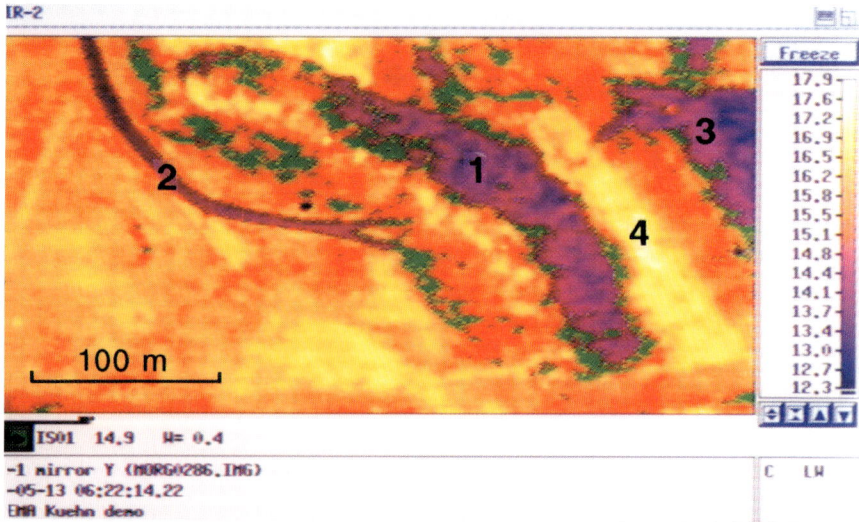

Fig. 6.11. Thermal image taken at 5:22 CET on May 13, 1993, of part of the Schoeneiche waste disposal site: (*1*) fresh waste, (*2*) a road, (*3*) synthetic foam cover (Piasol), and (*4*) a slope. Green: 14.9 °C isotherm (image: F. Boeker and F. Kuehn, BGR)

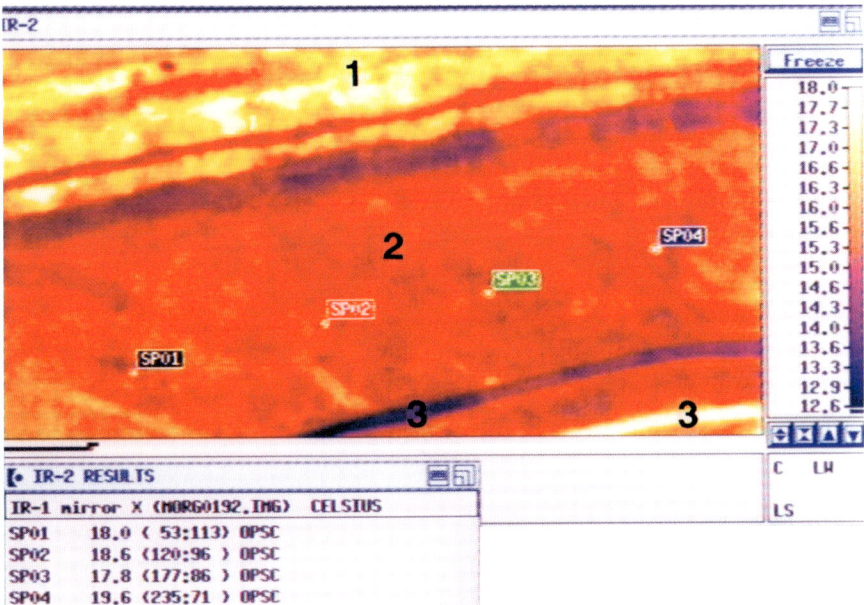

Fig. 6.12. Thermal image made at 5:15 CET on May 13, 1993, shows four emerging ducts (*SP01–04*) used for the dispersal of biogases. "*AGEMA 900*" software for image processing and evaluation permits calculation of the "radiation temperature" at the gas ducts, which fluctuates between 17.8 and 19.6 °C. The temperature rise of 1.8 °C between SP03 and SP04 indicates a temperature increase within the landfill. (*1*) Slope, (*2*) covered waste, and (*3*) roads (image: F. Boeker and F. Kuehn, BGR)

Fig. 6.11 shows (*1*) recent waste, (*2*) a road, and (*3*) synthetic foam cover (Piasol), which show almost identical temperatures. This image might look markedly different if it were acquired under different conditions.

Owing to the heterogeneous nature of the surface material, it is generally difficult to confidently detect relatively small temperature increases resulting from decay within household waste. Buried heat sources in a landfill can sometimes be detected if they are sufficiently hot. An example of an identifiable anomaly caused by smoldering pyrite under several meters of cover is shown in Fig. 4.4b.

Experience has shown that thermal images best provide useful information when employed to answer a specific, well-defined question. The best chances are for locating dumped materials concealed by a thin, homogeneous cover (Fig. 6.20 and Section 6.4).

6.3.2.2
Examining the Subsurface of a Waste Disposal Site

A series of aerial photographs taken between 1945 and 1976, prior to the development of the Schoeneiche waste disposal site was used to determine baseline information on the characteristics and state of the uppermost soil layers. The aerial photographs were analyzed to determine

- surface characteristics prior to landfill operations,
- areal distribution of impermeable and permeable soils (to determine potential areas for infiltration of leachate),
- natural drainage systems,
- former waterlogged areas below the present landfill,
- artificial drainage systems that provide pathways from the waste disposal site into the surrounding area.

Criteria for recognizing these characteristics can be illustrated by selected photographs. In an aerial photograph taken May 22, 1953, (Fig. 6.13) agricultural use is discernible by the field patterns in the entire area now covered by approximately 10–15 m of waste. Lighter colored patches in the individual fields are often thought to be either slight topographic highs or areas of sandy soil. However, stereoscopic evaluation shows no direct correlation between the light-colored areas and slight topographic highs; in fact, some of these areas were determined to be slight depressions. Thus, these light-colored areas are believed to have a high permeability (low soil moisture, well drained). Leachate from within the waste that now lies over these patches may be expected to drain directly into the uppermost aquifer through this permeable material.

Further inspection of the aerial photograph reveals that the occurrence of light-colored patches interpreted as areas of low soil moisture decreases and the number of drainage ditches increases toward the southwestern part of the site. Consequently, the southwestern part of the site is believed to be underlain by rather impermeable soil or have a shallow depth to the water table below the base of the landfill. Thus, both of the following cases are possible: either leachate cannot drain directly into the underlying aquifer or there is hydraulic contact between the lowermost waste layers and the unconfined aquifer. Clarification must be left to ground-based investigations.

It is often a coincidence if archival aerial photographs can contribute to clarifying site-specific problems. Seasonal conditions and the weather at the time the aerial photographs were taken determine their usability. Weather conditions during a reconnaissance flight in the Mittenwalde area on April 10, 1945, were optimal (Fig. 6.14). New vegetation growth patterns delineated substrate and moisture conditions in the area at the time the data was acquired.

At the present time, the area of faint stripes recognizable in Fig. 6.14 is covered by the northern part of the Schoeneiche landfill. The stripes may indicate long-term agricultural use (old field patterns) or the existence of abandoned drainage systems (see Schneider 1974). Although no old drainage systems were documented in this area, it cannot be excluded that the faint linear features (*arrow*) indicate an old artificial drainage system. A ground check of such features is recommended if they are observed within the area of a landfill. An abandoned drainage system may still function and may transport waste leachate away from the landfill. In this particular case, damaged plants and slightly discolored water were observed in CIR aerial photographs in which the feature joins the Gallun Canal (cf. Fig. 6.22).

Fig. 6.13. Aerial photograph taken on May 22, 1953, of cultivated land now occupied by the Schoeneiche landfill (*outline shown*). Light-colored patches (e.g., *arrow*) within the fields suggest that differences in soil substrates influence seepage at the base of the present site. N–S-oriented drainage ditches (*D*) in the SW part of the site are visible stereoscopically and indicate a shallow water table or low permeability of the soil (source: uve GmbH Berlin)

As mentioned above, archival aerial photographs were rarely taken for the reason that they are currently being used. It is pure chance that the time of day and season of the year are optimal for identification of moist areas at the margins of

Fig. 6.14. Wartime aerial photograph taken on April 10, 1945, showing cultivated land now covered by the northern part of the Schoeneiche landfill (*within dashed line*). The high contrast reveals differences in moisture content and lithology. Faint stripes (*arrow*) suggest the presence of an old drainage system or long-term agricultural use (old field patterns) (courtesy of Luftbilddatenbank in Würzburg)

a landfill. In Fig. 6.15, aerial photographs from 1992 and 1945 of a portion of the western margin of the Schoeneiche site are juxtaposed with a thermal image from 1993. Contrast is very low in the aerial photograph from May 1992 (6.15a) owing to homogeneous plant growth in the area; moisture-related features cannot be detected. In contrast, the aerial photograph from April 1945 of the same area has excellent contrast, and indicates areas of naturally elevated soil moisture below and around the present landfill (dark patches, 6.15b). A significant feature of elevated soil moisture can be observed in the thermal image

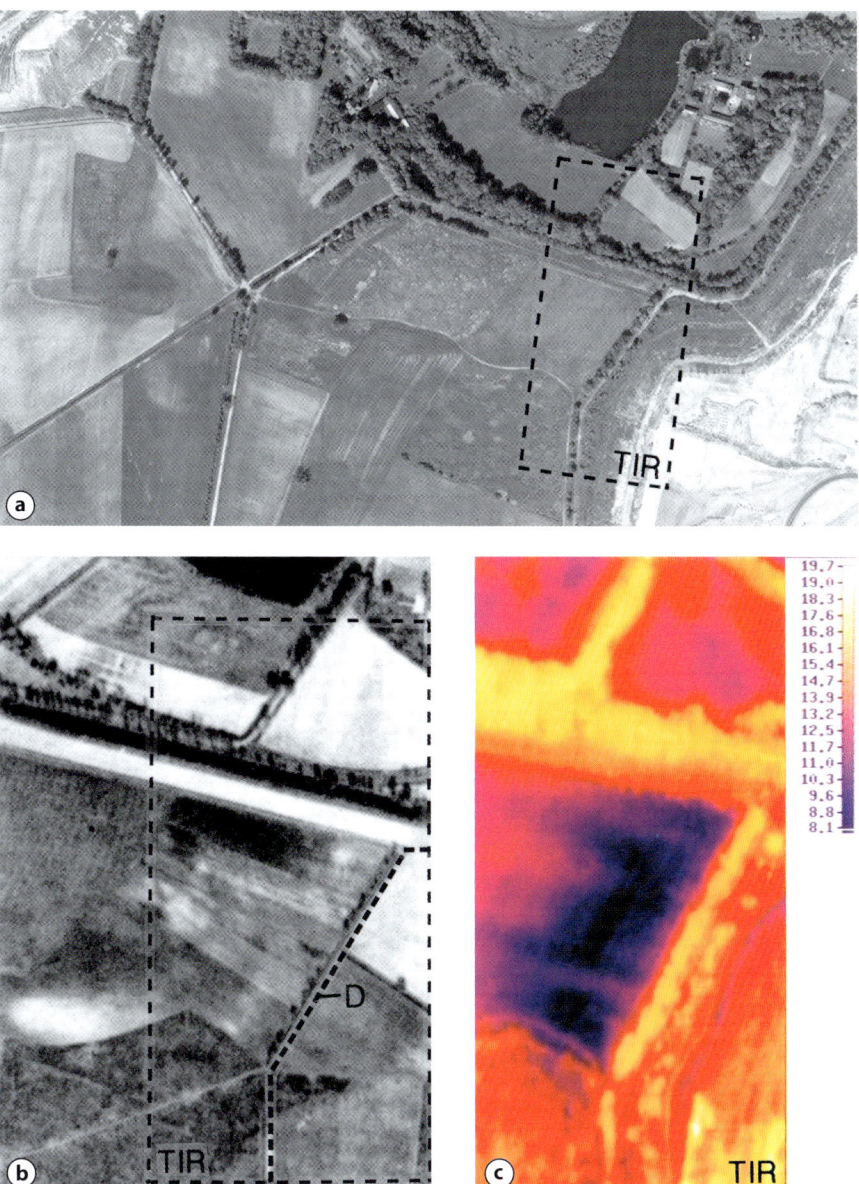

Fig. 6.15. Example of how the season and weather conditions prior to a flight influence the detail visible in an aerial photograph. **a** Aerial photograph taken on May 15, 1992, of part of the western margin of the Schoeneiche landfill. Details of the ground surface normally shown by different gray shades are masked by the homogeneous vegetation cover. **b** Aerial photograph taken on April 10, 1945 (*dashed line D* marks edge of the landfill); and **c** thermal image (TIR) taken at 5:10 a.m. CET on May 13, 1993. Both the lower photograph and the thermal image show evidence for an inhomogeneous distribution of soil moisture (sources: Landesvermessungsamt Brandenburg; Luftbilddatenbank in Würzburg; and BGR, respectively)

Fig. 6.16. Three aerial photographs of the Schoeneicher Plan landfill taken on **a** April 10, 1945, **b** May 22, 1953 and **c** June 23, 1967, showing the gradual filling of the abandoned clay pit with waste (sources: Luftbilddatenbank in Würzburg; uve GmbH Berlin; and Federal Archives, Potsdam branch, respectively)

from May 1993 taken just west of the landfill (6.15c). The dark blue indicates a cold, moist surface. In this particular case, a shallow ground water table is suspected to be the reason for elevated soil moisture. Under certain seasonal conditions, the ground water table may be temporarily in hydraulic contact with the bottom layer of waste.

In contrast to the relatively new Schoeneicher landfill, the neighboring Schoeneicher Plan site (Fig. 6.16) was first used for waste disposal in the 1920s and continued into the 1970s. Many different types of waste were dumped in abandoned, water-filled clay pits and are now in hydraulic contact with the uppermost aquifer. Aerial photographs document when and where waste material was dumped. Thus, it is possible to distinguish between sites used for household and industrial refuse prior to 1945, sites used for military and wartime waste from 1945 until the early 1950s, and sites used for waste in the 1960s and 1970s. The constant presence of water in the pits may lead to elevated rates of solution, corrosion, and elevated risk of contamination through possible direct hydraulic contact with the uppermost aquifer. Consequently, evaluation of the Schoeneiche landfill, only 200 m from the Schoeneiche Plan site, must take into consideration the potential impact of the neighboring landfill (cf. Figs. 4.2 and 4.3).

6.3.2.3
Examining the Immediate Vicinity of a Landfill

In a characterization of the area surrounding a waste disposal site, regional structures, potential migration pathways for leachate and indications of environmental impacts are identified. A regional study should also identify areas in which pre-land-fill activities occurred that could add to the pollutance emanating from the landfill. This could be useful if responsibility for pollution in the area has to be addressed.

Evaluation of the Schoeneiche landfill and vicinity included the examination of aerial photographs and thermal images to evaluate

- areas with elevated soil moisture, springs, or areal discharge from the uppermost aquifer,
- changes in the character of the exposed soils,
- areas of suspected contamination that could affect soil and ground water,
- vitality of the vegetation.

Moist Areas and Springs

As discussed previously, moist areas and springs can be easily detected by remote sensing techniques. Moist areas commonly develop at the surface due to the presence of perched groundwater or to a shallow water table. Geochemical sampling of moist areas may identify the sources of the moisture by their geochemical signatures.

The archival aerial photograph in Fig. 6.17 shows an area about 100 m northwest of the Schoeneiche landfill. A system of ditches can be seen that are partially filled with silt (*D*), indicating that this area must be permanent wetland.

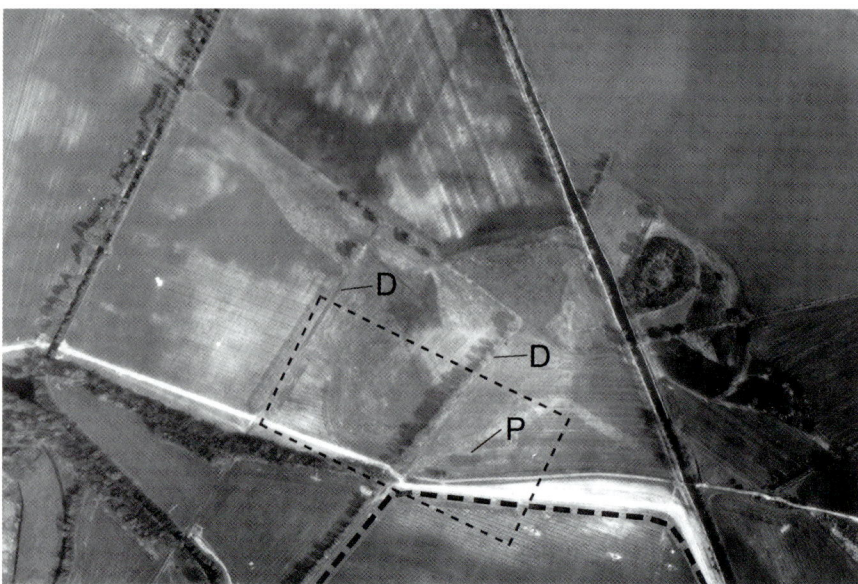

Fig. 6.17. Aerial photograph taken on April 20, 1974, of an area just north of the Schoeneiche landfill. The dark gray areas indicate elevated surface moisture content. The water has been drained from some areas via partially silted-up ditches (*D*). Outline of Fig. 6.18 (*thin dashed line*) and margin of the landfill (*thick dashed line*) are shown for reference. Dark patches (*P*) indicate an abandoned landfill (photo from the Federal Archives, Potsdam branch)

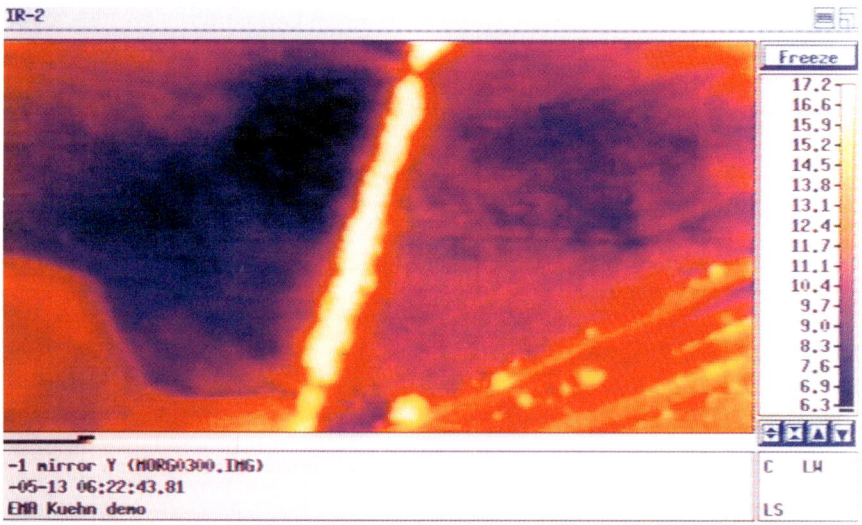

Fig. 6.18. Thermal image taken at 05:22 CET on May 13, 1993, of the area marked in Fig. 6.17, which shows low temperatures (*dark areas*), indicating a high soil moisture content (image: F. Boeker and F. Kuehn, BGR)

This assumption is supported by thermal images of the site (Fig. 6.18). Signs of elevated water content can be seen in the thermal images. An extensive "cold" (*dark blue*) zone indicates a high moisture content in the topsoil. Electrical conductivities around 1 200 µS cm^{-1}, or about four times the normal value, were measured in groundwater samples from this zone suggesting that mineralized groundwater containing leachate from the landfill rises about 200 hundred meters north of the waste site.

Dark patches (*P*) in the the aerial photograph (Fig. 6.17) also suggest surface inhomogeneity, which a ground check showed to be due an abandoned landfill. The extent to which the abandoned landfill contributes to the elevated ground water mineralization has to be checked.

Abandoned Landfills and Areas of Suspected Contamination

When soil and ground water contamination is detected, questions concerning the temporal and spatial dispersal of contaminants arise. Because remediation can be expensive, and the polluter is legally required to bear the cost, the landfill operator will want to establish the actual sources of the contamination to avoid charges of clandestine waste disposal that often occurs at large sites.

In 1992, the Working Group for Remediation and Management of Waste Disposal, Berlin, (*Arbeitsgemeinschaft "Sanierungs- und Betriebskonzept / Abfallentsorgung Berlin"*) conducted a survey of abandoned waste disposal sites and other areas of suspected contamination in the vicinity of the Schoeneiche and Schoeneicher Plan landfills. A number of operating and abandoned waste disposal sites, as well as sewage and waste water irrigation systems, were detected and mapped.

The following example demonstrates how aerial photographs and thermal images can be used to detect and characterize an abandoned landfill. An elevated area typical of landfills can be identified between the Schoeneiche and Schoeneicher Plan sites in the center of the stereo-pair photographs in Fig. 6.19. The area is immediately southeast of the water-filled pit (*black*). It shows patterns typical of cropland. Additional information was provided by thermal data (Fig. 6.20). The thermal image reveals temperature anomaly patterns oriented more or less N–S in this area, most likely indicating near-surface changes of material (*dark blue and red patches*). NW–SE-oriented yellowish features are due to the grain crop. The image helped to define areas for taking samples to check the character of materials suspected of being dumped in the area.

Soil Characteristics

Information about the characteristics of the soils in the vicinity of the landfill is required to identify possible migration pathways for leachate from the landfill. Figure 4.8 demonstrates how CIR aerial photographs can be used to differentiate between sandy (i.e., permeable) and cohesive soil. This CIR aerial photograph, taken on July 2, 1993, shows farmland just west of the Schoeneicher Plan landfill with a 15-cm-high grain crop. It illustrates the correlation between soil characteristics and plant vitality. The red areas denote dense, advanced plant growth due to high nutrient availability and/or elevated soil water content. The light-colored ar-

Fig. 6.19. Portion of a CIR aerial photograph taken on July 3, 1993, (*stereopair*) showing an abandoned landfill (area in the center just below the body of water, *arrow*) between the Schoeneiche and Schoeneicher Plan landfills (photo taken by WIB GmbH Berlin for BGR)

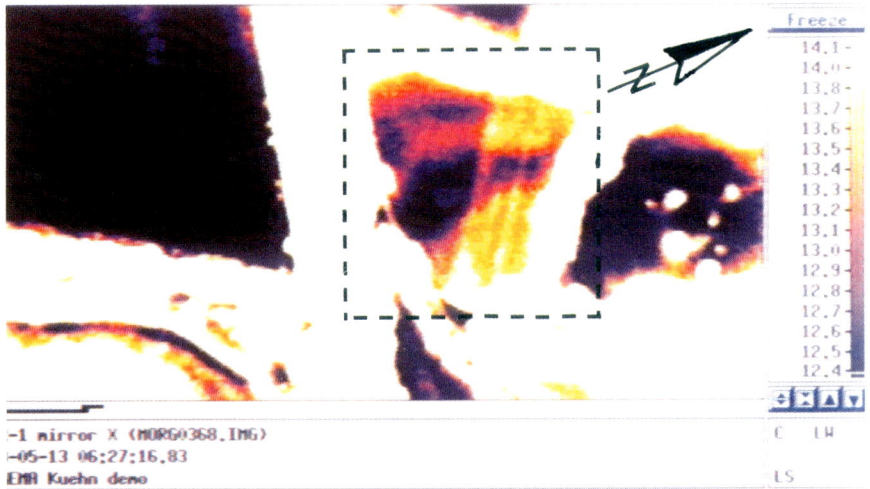

Fig. 6.20. Thermal image taken at 05:27 CET on May 13, 1993, showing the landfill (within *dashed line*) in Fig. 6.19 and temperature anomalies resulting from a partial vegetation cover (*yellow and orange stripes*) and from soil (*dark blue, stripes* trending roughly N–S); (taken by F. Boeker and F. Kuehn, BGR)

eas with less healthy plant growth indicate lower nutrient availability and/or permeable near surface beds. Ground checks confirmed the presence of sandy soils in these areas. The sandy areas have a high permeability, facilitating seepage from the adjoining landfill into the unconfined, uppermost aquifer.

Vitality of Trees

CIR aerial photographs have been used to evaluate the vitality of vegetation. CIR film is sensitive to the visible and near infrared portions of the electromagnetic spectrum (NIR-I, see Table 2.1 and Section 3.2.1). The NIR-I data provides an indication of the health of vegetation that would otherwise not be observable with the human eye. Besides natural causes due to the soil and subsoil, unhealthy vegetation may be caused by anthropogenic effects. When trees or other plants are used as bioindicators of potential soil or ground-water contamination, it must be determined whether the plants have direct root contact with the contamination.

CIR aerial photographs are effective tools for analysis of vegetation vitality, particularly in areas such as the Schoeneiche area, where a shallow aquifer is present. Under these conditions, root contact with the uppermost aquifer is likely.

CIR aerial photographs used to delineate areas of plant damage should be acquired in the second half of the growing season. The second half of the growing season avoids both the springtime vegetation burst, when small or moderate changes in vitality are hardly detectable, and the beginning of the seasonal senescence associated with autumn. Experience has shown that the optimum scale to obtain the data is between 1 : 2 000 and 1 : 10 000; photo scales of less than 1 : 10 000 do not have the spatial resolution necessary for most site studies.

As discussed in Section 3.2.1, the red color of healthy vegetation in CIR aerial photographs results from the strong reflection of near-infrared radiation by leaves. Bright red colors in CIR aerial photographs are generally associated with the presence of healthy grass, broad-leaved plants. In contrast, healthy conifers are a much darker red in these false-color images (see Fig. 6.8, *bottom*). In general, the decomposition of chlorophyll, which is related to changes in the leaf structure, causes NIR-I radiation to be reflected less intensely. Thus, less healthy leaves and needles will loose their red color in the photographs proportionally to the degree of damage. Although some trees show different color changes in response to stress (e.g., damaged oaks and willows turn black; birches, alders, and willows turn brown; linden trees (*Tilia*) turn bluish-green, and silver willows become whiter). In addition, a decrease in tree vitality is recognizable by the typical thinning of the crown. Using these criteria, a 4- to 5-step scale can be developed to visually assess the vitality of vegetation using CIR aerial photographs.

Trees, because of their long lifetime and deep root penetration, are useful indicators of many types of environmental impacts. Prior to mapping the vitality of trees, it is necessary to be able to distinguish the different tree species because each species shows characteristic variations of the red color. Attributing differences in leaf color in the photographs simply to environmental factors can lead to serious misinterpretation. A color key for photo interpretation developed immediately after taking the CIR aerial photographs can eliminate potential errors due to seasonal or weather-related changes in vitality.

In principle, the major tree species in the area around the Schoeneiche landfill can be identified on the basis of their healthy species-specific color in the CIR photographs as follows:

- oak   bright red,
- silver willow silvery red,
- pine   red-brown.

Figure 6.21 shows a CIR aerial photograph taken on July 2, 1993. The image shows the area immediately north of the Schoeneiche landfill and several rows of trees

Fig. 6.21. CIR aerial photograph taken on July 2, 1993, showing the area just north of the Schoeneicher Plan landfill. The road along the north edge of the landfill is just visible at the *bottom right* (photo taken by WIB GmbH Berlin for the BGR)

that grow along both the open and silt-filled ditches (cf. Fig. 6.17). The trees are primarily willows, black poplars, and alders.

A map depicting the health of trees in the vicinity of the Schoeneicher Plan and Schoeneiche landfills is shown in Fig. 6.22. The map was compiled using only CIR aerial photographs from July 2, 1993, and ground-check information. This map supports the earlier assumption that contamination spreads northward from the landfill. It can also be seen that the row of black poplars northeast of the landfill along the Galluner Canal are considerably stressed, supporting the suggestion that seepage water from the landfill drains via an old drainage system. A potential impact of the neighboring waste incineration plant on the health of the black poplars cannot be ignored, however.

6.3.3
Summary

Remote-sensing investigations were used to characterize materials and geology on local and regional scales at the Schoeneiche waste disposal site. The investigations described here were supported by ground checks, including geochemical and geophysical analyses, that verify the interpretations. The northern part of the Schoeneiche landfill and vicinity is shown in Fig. 6.23, mapped using remote sensing data.

Evidence of an old abandoned waste dump (confirmed in the meantime by drilling and trenching) was found in the area just north of the landfill. Seepage from the uppermost unconfined aquifer and willows showing considerable stress were observed in the area. The ground water was found to have an elevated electrical conductivity (see discussion of Figs. 6.17 and 6.18). The question remains, however, as to whether the dissolved contaminants are from the abandoned waste dump, or from the present Schoeneiche site or both; additional hydrochemical analysis is needed.

Areas of primarily sandy soil where leachate and precipitation rapidly infiltrates into the uppermost aquifer are indicated in Fig. 6.23. Suspected old drainage systems are also mapped because they represent potential drainage pathways for seepage at the base of the waste.

Evaluation of the thermal images yielded only supplementary information on the waste itself. Low-level heat sources, such as those generated by decay in landfills for household waste, were difficult to detect.

In landfills and mine dumps containing high-temperature heat sources, thermal remote sensing has been much more effective, particularly for shallow sources (see example discussed in Section 4.2.1).

The use of thermal remote sensing for the area around the Schoeneiche and Schoeneicher Plan landfills yielded very good results in localizing abandoned waste dumps beneath farmland, moist patches, and areas of permeable soil (see Figs. 4.15 and 6.20). However, similar positive results should not be expected for all areas.

In this case study, remote sensing aided the assessment of the general setting of the site, characterization of the base of the landfill, and detection of impacts in the surrounding area. This knowledge helped effectively to apply subsequent ground-based methods (geophysics, drilling, soil and water sampling, etc.).

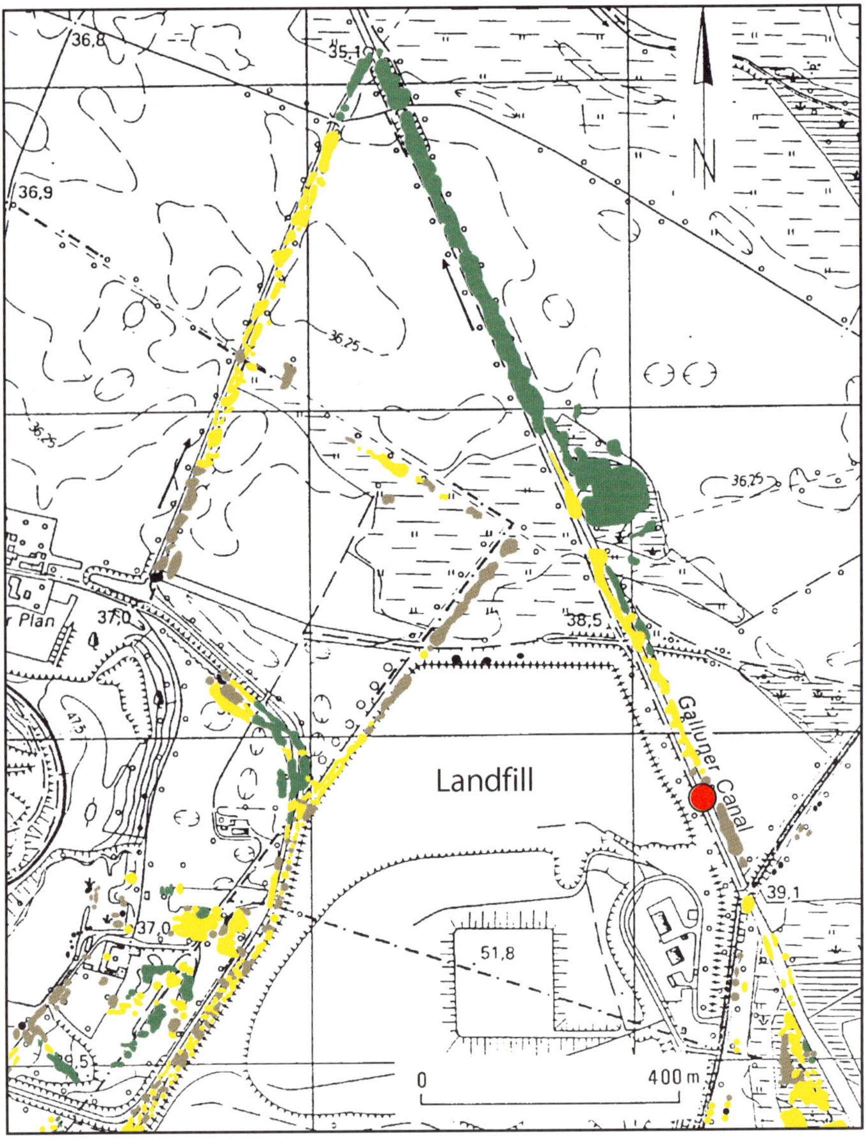

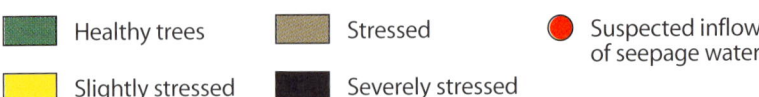

Legend

Topography after topographical survey sheet Tk 10

- ■ Healthy trees
- ■ Slightly stressed
- ■ Stressed
- ■ Severely stressed
- ● Suspected inflow of seepage water

Fig. 6.22. Map of the vitality of trees in the vicinity of the Schoeneicher Plan and Schoeneiche landfills; mapped at a scale of 1:5000 (simplified)

CHAPTER 6 · Case Studies

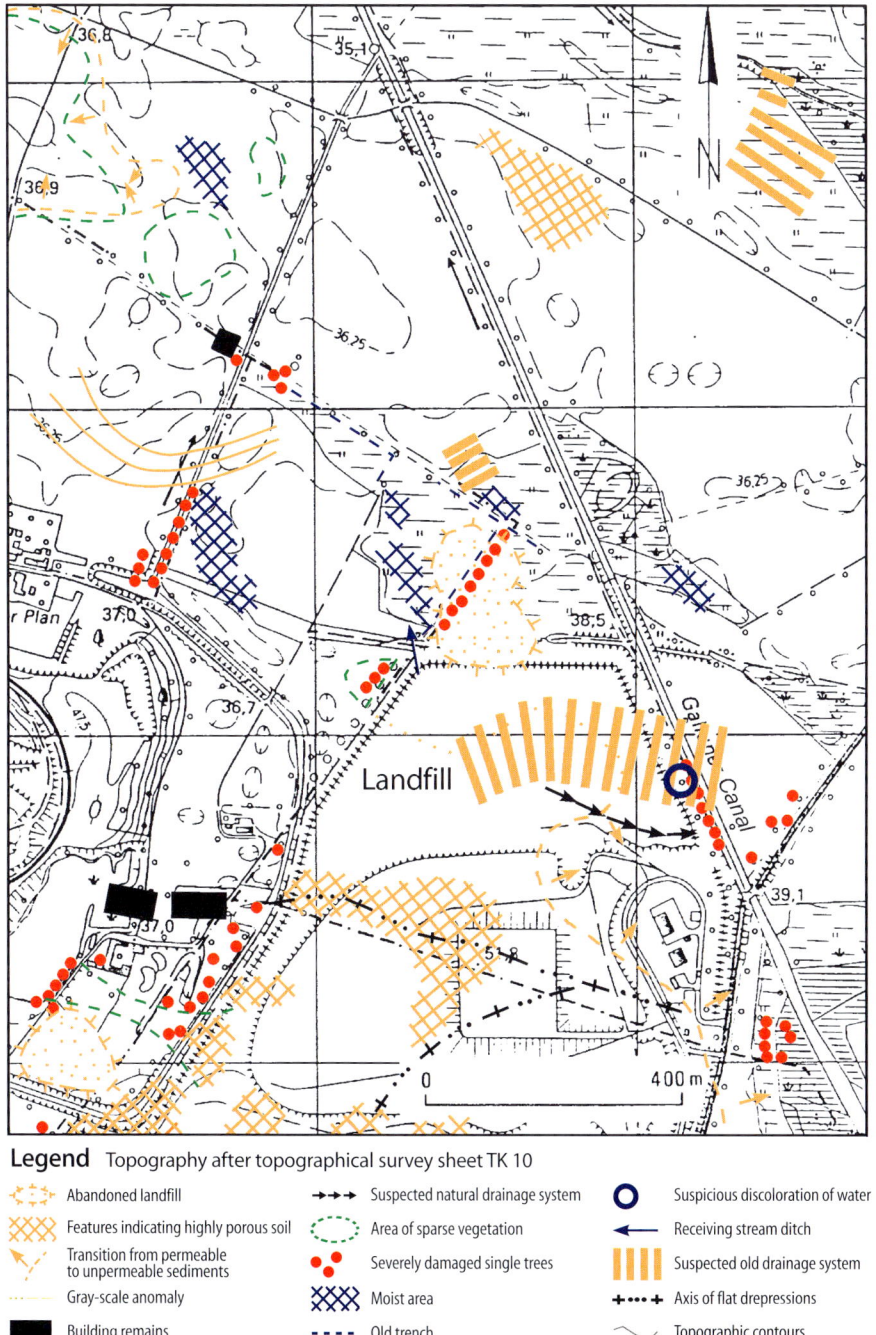

Fig. 6.23. Map of the same area as in Fig. 6.22. Interpretation of remote sensing data for the area around the northern part of the Schoeneiche landfill (simplified)

6.4
Thermal Remote Sensing to Detect Buried Waste Material (Oak Ridge, U.S.A.)

John M. Irvine · Gary Stahl · Julie Odenweller · John L. Smyre · Thomas K. Evers
Dale Huff · Amy L. King

6.4.1
Introduction

The legacy of America's nuclear weapons and nuclear research programs is extensive. The Department of Energy (DOE) operates numerous waste sites containing high-level radioactive waste, low-level radioactive waste, hazardous chemical waste, solvents, contaminated machinery, high explosives, and noncontaminated construction debris. The monitoring and remediation of these sites is expected to consume substantial resources over the next several decades. In-situ measurement and analysis of these sites is both expensive and potentially dangerous. The use of remote sensing can provide a cost-effective source of information, while minimizing human exposure to environmental risks. Although remote sensing can not fully replace the need for direct ground-level observation at these sites, it can significantly reduce the requirements and suggest effective and safe sampling strategies for collecting the in-situ measurements.

This study addresses the detection and "geolocation" of buried waste trenches at known waste sites; locating previously undocumented waste sites is a separate issue not addressed here. Burial of hazardous materials has been a common practice since the early days of U.S. nuclear programs. Each of the major DOE laboratories has multiple areas containing known burial sites. The records, however, often provide only limited information about the precise trench locations and the contents of these trenches. In some cases, the trenches' contents can be inferred from site records and/or sampling of contaminants in the vicinity.

The environmental risk posed by these trenches depends on the contents and the tendency for contaminants to migrate into the surrounding environment. Materials which are well-contained and have short half-lives, for example, are of less concern than trenches that are "leaking" dangerous contaminants. A variety of techniques are available for remediating these buried waste sites. Precise knowledge of the trench location and boundaries may permit the remedial action to focus on specific problem trenches, rather than the entire waste site. This more focused approach can result in significant resource savings.

6.4.2
Background

Waste Area Group (WAG) 4, one of 17 WAGs associated with Oak Ridge National Laboratory (ORNL), is located along Lagoon Road south of the main plant complex. Solid Waste Storage Area (SWSA) 4 is the largest unit at WAG 4, covering approximately 23 acres (see Fig. 6.24). In the 1950s, SWSA 4 received a variety of low- and higher-activity-level wastes, including transuranic wastes, all buried in trenches

Fig. 6.24. Oblique aerial photograph taken from a helicopter shows SWSA 4 today. Note that the grassy field offers limited and ambiguous indications of the waste buried below

or auger holes. During the period 1955 through 1963, SWSA 4 was designated the Southern Regional Burial Ground for the Atomic Energy Commission (predecessor to the DOE). Approximately half of the waste received at SWSA 4 in the 1950s originated at ORNL, while the remainder came from a number of off-site locations (Oak Ridge Y-12 Weapons Plant, Argonne National Laboratory, Knolls Atomic Power Laboratory, Mound Laboratory, and the General Electric Company). SWSA 4 received nonradioactively-contaminated construction debris from 1959 until 1973.

Identification and location of the waste trenches is important due to the contaminants released from SWSA 4. Surface water sampling data indicate that a significant quantity of contamination is being released from these trenches. Contamination released from the trenches accounts for 25% of the strontium-90 observed at White Oak Dam in the 1987–1994 period and about 14% of the total ORNL off-site human health risk via the drinking water pathway. Additional contaminants of concern include cesium-137. Stemming the flow of these contaminants requires identifying the trenches that contribute the majority of these contaminants and undertaking remedial actions to contain and isolate the buried waste from surface and subsurface water.

A fire in 1957 destroyed the waste disposal records for SWSA 4. The primary surviving information about trench locations is a line map constructed from interviews with site workers (see Fig. 6.25). Field investigations at SWSA 4 have con-

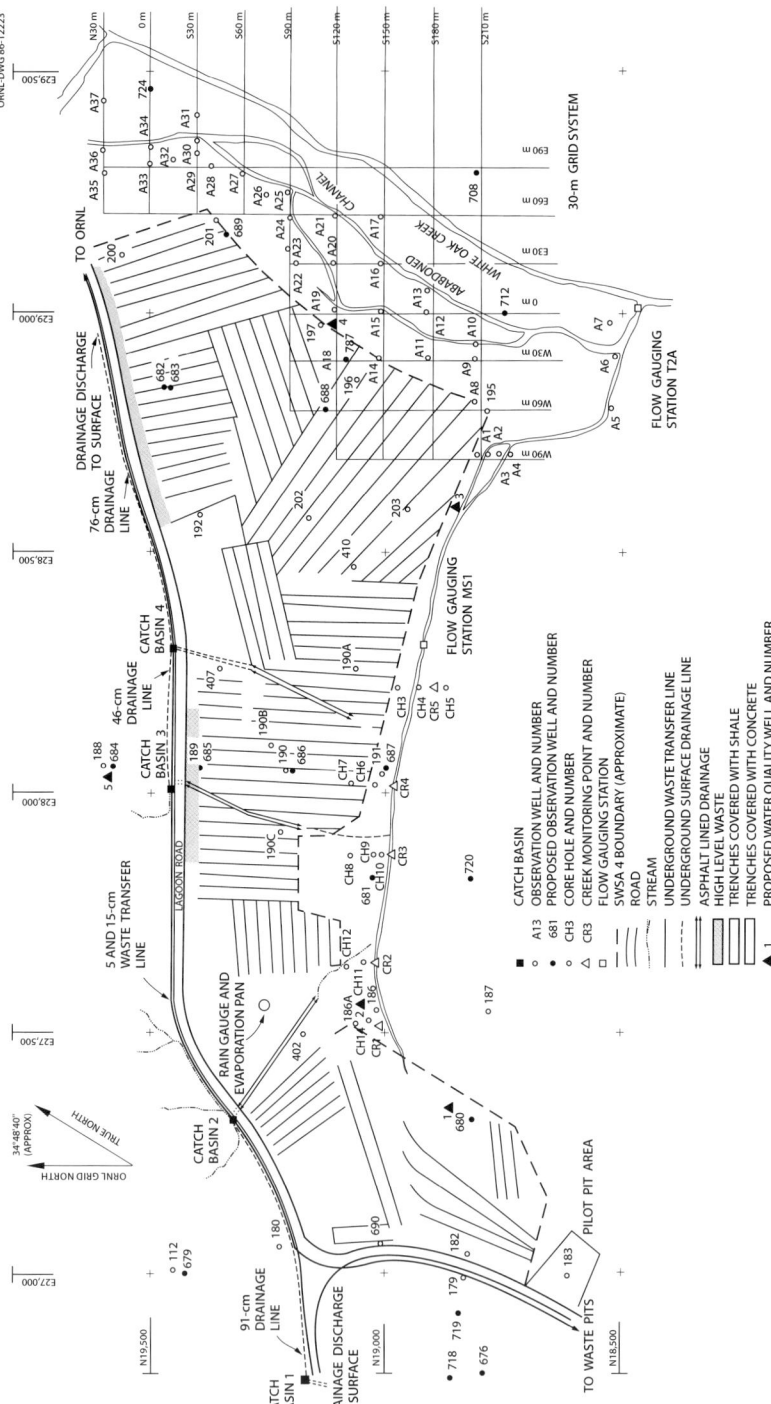

Fig. 6.25. Historical trench map constructed from interviews with site workers following a fire that destroyed many of the records for SWSA 4. Although the map correctly depicts the general orientation of the trenches, trench locations are not at all accurate

firmed the general orientation of the trenches depicted on the line map, but could not determine the precise location or orientation of specific trenches.

ORNL is characterized by a succession of elongated and relatively parallel southwest-trending valleys and ridges, formed by the folded limestone structures underlying the region. The soils are derived from underlying elastic and carbonate bedrock, and typically consist of red yellow and red-brown silts and clays. The soil at the site is moist, low in organic content, strongly leached, and acidic (pH 4.5 to 5.7). Illite and vermiculite (clay materials) in the soil enhances the soil's ability to retard migration of contaminants through matrix diffusion, mechanical filtering absorption, and ion exchange.

6.4.3
Imagery Analysis

The study used both historical aerial photography and recent thermal and multispectral imagery of the Oak Ridge Reservation (ORR). The historical imagery, spanning 14 collections, was obtained from the U.S. Department of Agriculture, the National Aerial Photography Program, and the Tennessee Valley Authority. All data are black-and-white panchromatic photography. The scale varied from 1 : 12 000 to 1 : 43 200. All archival data were available only in hardcopy, either as photographic prints or film transparencies.

High resolution multispectral data from the *Daedalus 1268 Multispectral Scanner* were available over SWSA 4. The *Daedalus* scanner, which has frequently been used over Oak Ridge and other DOE sites, has twelve spectral bands spanning the near ultraviolet, visible, near-infrared, short-wave infrared, and long-wave (ther-

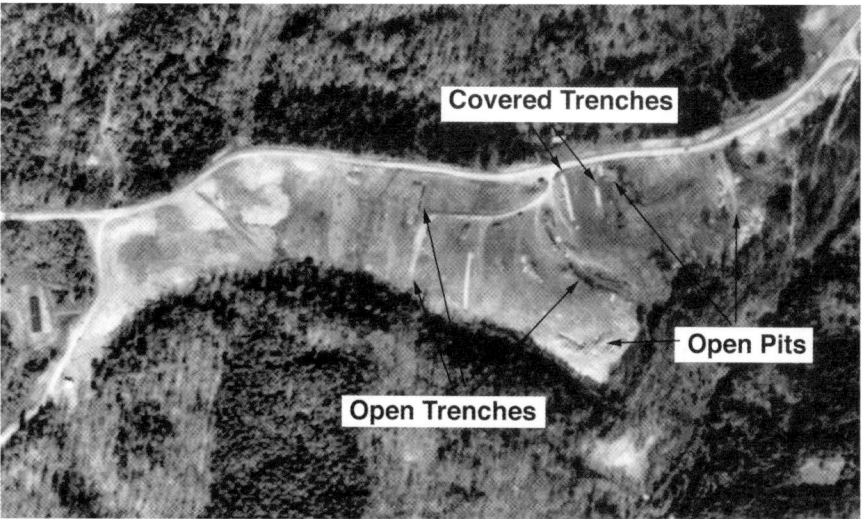

Fig. 6.26. New and recent waste trenches are evident from stereoscopic analysis of this early historical photograph, which was collected on 29 Feb. 1956 during the period of waste burial at the site

mal) infrared wavelengths. The spatial resolution varies from one to five meters GSD, depending on the altitude of the collection flight.

Locating the trenches on historical imagery required stereoscopic analysis of the hardcopy photos. Trench locations were then located on digitized softcopy imagery and registered to a common coordinate system. The images were then rectified and registered to the administrative grid, which is the local Oak Ridge coordinate system. Registration of the imagery to the Oak Ridge Administrative Grid, which is the local coordinate system, was difficult due to the shortage of fixed reference points in the area. Lagoon Road appears in all of the images, but there is a lack of fixed points along the southern boundary of SWSA 4.

The collection of archival photos permitted direct observation of the burial activities in the 1950s. Although some gaps in coverage exist, open trenches are apparent on photos from this area (see Fig. 6.26). In addition, texture difference and subsidence features indicate the presence of trenches that were recently closed (see Fig. 6.27). Trench locations identified from archival photos correspond to signatures evident in the Daedalus data, as well. The *Daedalus* thermal channel (8.5–12.5 μm) collected during daytime hours shows an apparent thermal difference between the trenches and the surrounding areas (see Fig. 6.28). This signature corresponds to the temperature and soil moisture differences (low temperature, high soil moisture) observed at the site through ground measurements discussed in the next section. In the false-color composite image, the vegetation over the trench locations exhibits greater vigor (see Fig. 6.29).

Each of these sets of imagery contributes information about the trench locations, but no single image or photo provides all of the requisite detail. Initially, information from the images was collated by using one image as the reference and

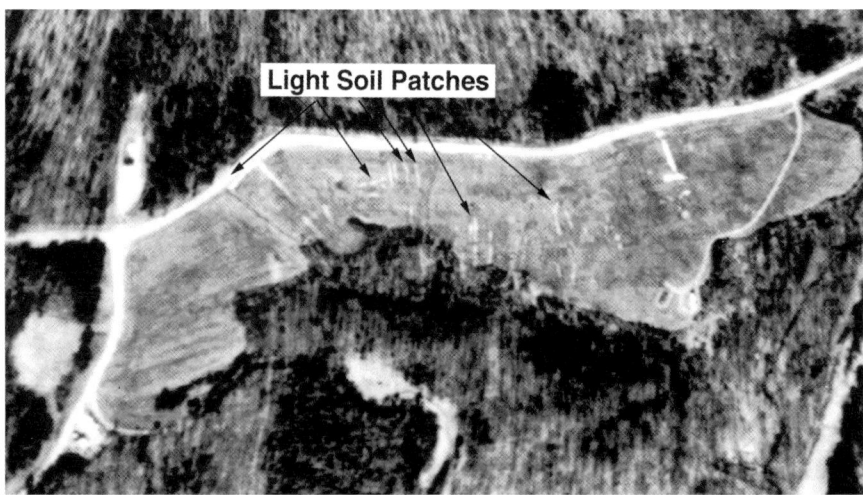

Fig. 6.27. This historical photograph was acquired on 25 March 1981, after the site had been closed to waste burial. Due to subsidence and the addition of fill soil, trenches are evident in this image as light toned patches

registering the other images to that reference. To develop a map of the trench locations, it was necessary to warp and register all of the information to the administrative grid. When this was done, individual trenches could be identified and compared across the images. Using this approach, a composite trench map was constructed (see Fig. 6.30).

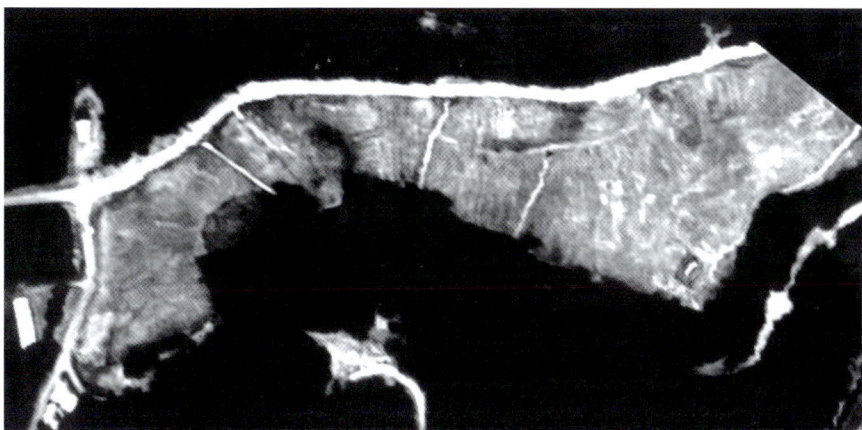

Fig. 6.28. Daytime longwave thermal image (8.5–12.5 μm) from *Daedalus* data, shows darker features (cold) corresponding to the trenches. Ground temperature measurements suggest that the signature would be more pronounced in predawn imagery

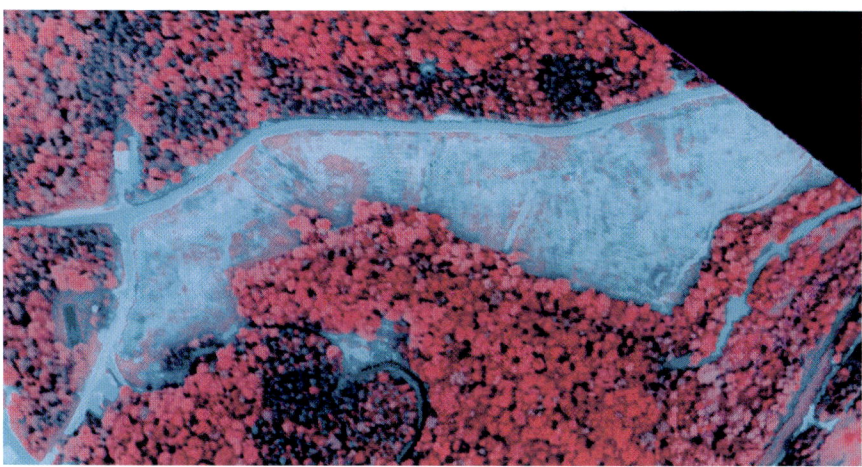

Fig. 6.29. This false-color composite multispectral image was formed from the near IR, red, and green bands (bands 7, 5, and 4 from the *Daedalus 1268 multispectral scanner*, cf. Table 3.2) placed in the red, green, and blue bands, respectively. Differences in vegetation vigor between the trench (red) and nontrench (blue) areas are apparent

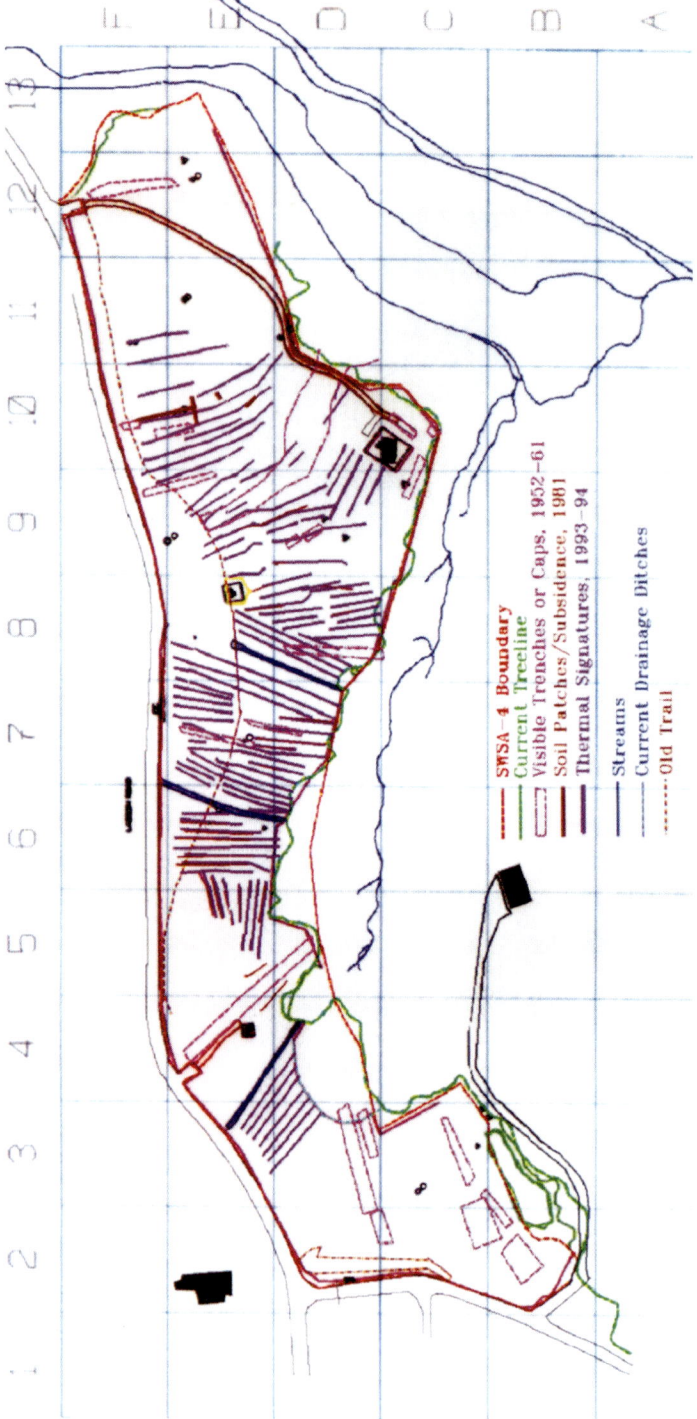

Fig. 6.30. Imagery-based trench map produced by collating the information available from all the imagery and registering it to a common reference

6.4.4
Ground Data

An array of ground sensors was set up at SWSA 4 to collect data on the surface and near-surface conditions in the trench, in an adjacent "nontrench" area, and at a nearby control location. The "nontrench" area is at the upslope end of one trench and, in fact, may have been partially disturbed during the trench excavation; however, no actual waste is buried in this nontrench area. The control area was used as a representation of an undisturbed soil. The data collected at this site included:

Soil Temperature. Measurements of the soil temperature were recorded using multiple thermistors at approximately 2.5-cm depth in each of the three areas. Additional thermistors were arrayed in two vertical profiles, at depths of 10, 15, and 25 cm, to assess soil temperature at greater depths. A total of 24 thermistors were employed in the data collection.

Thermal Radiance. Two IR transducers, one over the trench and one over the control area, recorded emitted energy in the longwave IR (8–14 μm).

Soil Moisture. An array of 10 instruments recorded soil potential, which is inversely related to soil moisture, at 5-cm depths both inside and outside the trench.

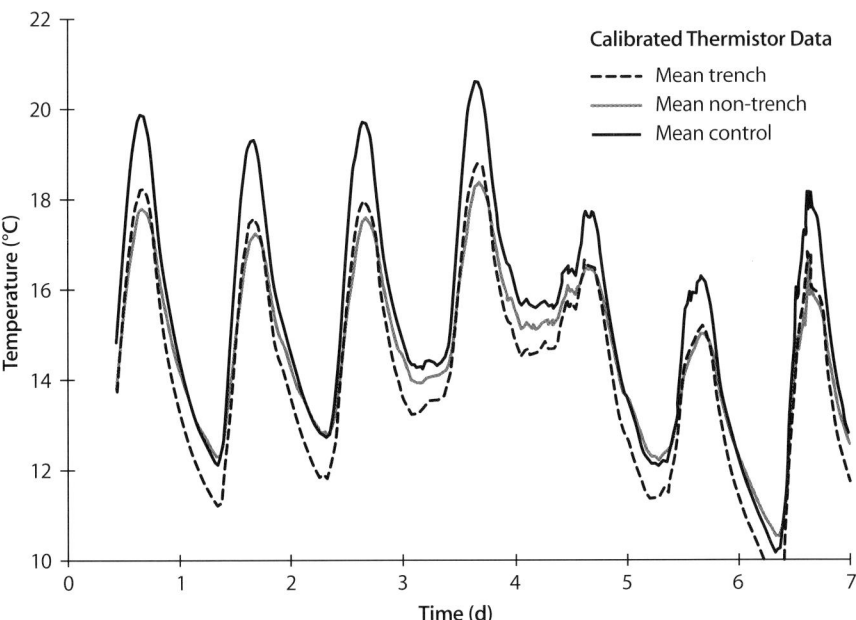

Fig. 6.31. Surface soil temperature measurements from multiple thermistors were averaged to form the plot shown here (calibrated termistor data). Diurnal variation in temperature is the dominant effect, but the differences between the trench and control areas are also clear

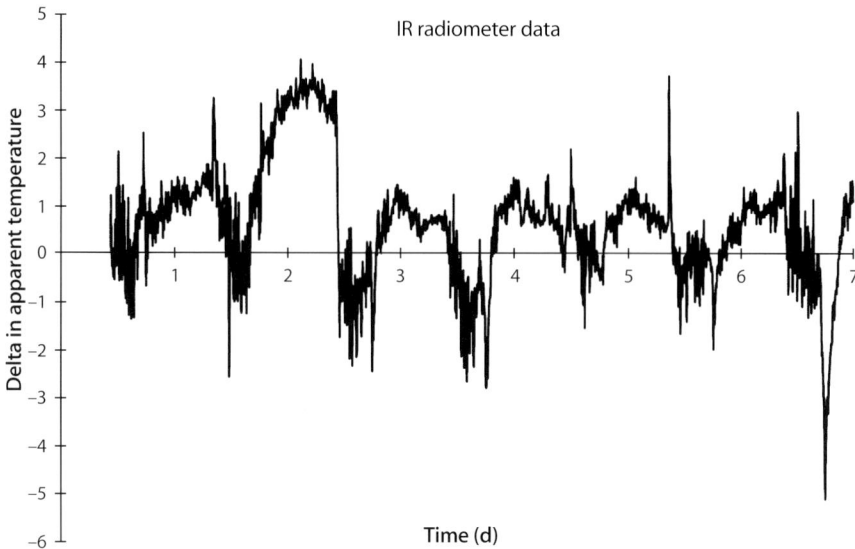

Fig. 6.32. Thermal radiance difference between the two IR transducers (one over the trench and one over the nontrench area) is plotted as a function of time. Note the clear nonzero values corresponding to nighttime, while the daytime differences are often ambiguous

Statistical analysis of the ground data indicates the nature of the thermal signature associated with the trenches. Trenches are typically cooler than the control area through the diurnal cycle (Fig. 6.31).

The nontrench area, which is located at the uphill end of the trench, varies over the diurnal cycle. If this nontrench area were partially disturbed during the initial excavation of the trenches, this could account for the thermal behavior. Both the thermistor and radiometer data (see Fig. 6.32) show that the difference between the trench and control areas is most evident at night. In the daytime, these data indicate some amibuity or reversal in the relative temperature difference. By comparison, the termistor measurements show a thermal difference between the trench and non-trench during the day, although the difference is more pronounced at night. Soil moisture measurements show that the trench area generally exhibits greater soil moisture, which may account for the observed thermal differences (see Fig. 6.33). These data suggest that repeated predawn thermal imagery should provide the data necessary to accurately map the trenches.

6.4.5 Conclusions

The field investigations at SWSA 4 identified 6 individual seep areas and suggested that they contributed more than 90% of the strontium-90 release from the study area. The trench map derived from the remote sensing data was the key factor in

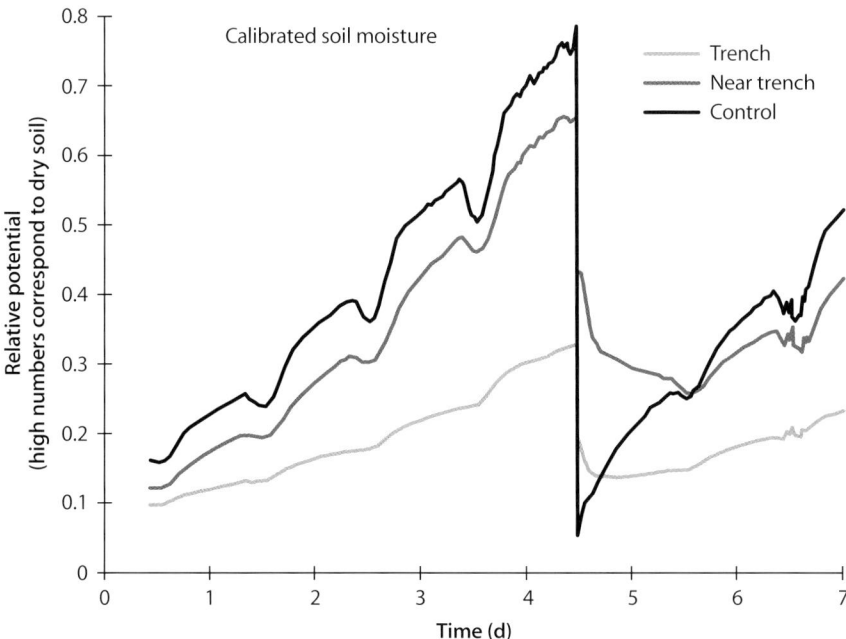

Fig. 6.33. Soil moisture data indicates a clear difference between the trench and nontrench areas. The differences increase over time, until a rain event saturates the area and then the divergence begins anew

pinpointing localized sources feeding the major seeps. Control of these few sources is the key to a cost-effective interim action to reduce strontium-90 releases and off-site risk.

Without the remote sensing results, the ability to quickly and effectively pinpoint the individual sources would have been lost. The alternative for controlling releases (i.e., cap the whole site, collect and treat surface water) would cost in excess of $5 million (US) more than directly controlling the sources. Thus, the use of remote sensing data provided an approach that could significantly reduce remediation costs for this site.

The procedures demonstrated here are applicable to numerous waste sites where potential contaminant migration from burial trenches or pits is a concern. Both DOE reservations and military facilities have a number of such sites. The alternative to employing remote sensing is to rely on extensive and costly ground sampling to precisely locate the buried materials. Although the use of remote sensing will not completely eliminate the need for ground observation, it can substantially reduce the amount of ground-level sampling required for site characterization and monitoring. Proper use of remotely sensed data can assist in locating trenches with greater precision, reduce the need for in-situ sampling, and ensure greater safety in the clean-up process.

6.5
Remote Sensing for Monitoring the Effects of Mining in Sudbury, Canada

Vernon Singhroy

6.5.1
Introduction

The deleterious effect of past hazardous waste is an international problem. In Canada and the United States alone there are over 100 000 abandoned mine sites. Our case study, in the Sudbury mining district, has shown that multispectral remote sensing techniques can detect the vegetation damage caused by the acid drainage from mine and mill tailings and waste rock and can monitor revegetation success at sites undergoing restoration.

Several remote sensing case studies on monitoring the effects of mine and mill tailings and acid mine drainage on vegetation have been conducted in Canada. The sensors used include airborne and spaceborne multispectral imaging systems and aerial videography. Graham et al. (1994) used principal component analysis techniques on *Landsat Thematic Mapper* images to monitor vegetation changes in large areas affected by iron ore mining operations at Noranda, Quebec. Mussokowski (1983) used classification techniques of multidate *Landsat TM* data to monitor vegetation change over a 10-year period in the Sudbury District. Similar techniques also were used by Hornsby et al. (1989) for placer gold mining areas in the Yukon. Digitized aerial photographs, integrated with topographic and drainage data, were used to characterize gold mining areas around Timmins, Ontario (Mussokowski et al. 1993). Airborne multispectral techniques are the most effective way to detect and monitor vegetation damage at mine sites and have been used successfully by Singhroy et.al. (1989), Singhroy and Kruse (1991), King (1993), and Singhroy (1995).

6.5.2
Sudbury Case Study

The Sudbury mining district is one of Canada's richest mining districts. This elliptical feature, known as the Sudbury Basin, is believed to have been formed by a meteorite impact nearly 2 billion years ago. Nickel and copper ore have been mined for over a century from more than 90 mines around the 150-km rim of the basin. The ore is derived from the sublayer of the Sudbury Igneous Complex, a fine to medium grained noritic to gabbroic rock containing as much as 60% sulfides. Ore grades sent to the mill are typically 1.2% Ni, 1.1% Cu, and 9.2% S. Figure 6.34 shows some regional geologic information and the distribution of mining properties of INCO, one of the largest mining companies in Canada.

6.5.2.1
Regional Environmental Restoration

The Sudbury mining district is one of the world largest metal smelting complexes and is a well known polluted region where the landscape has been devastated. The

Chapter 6 · Case Studies

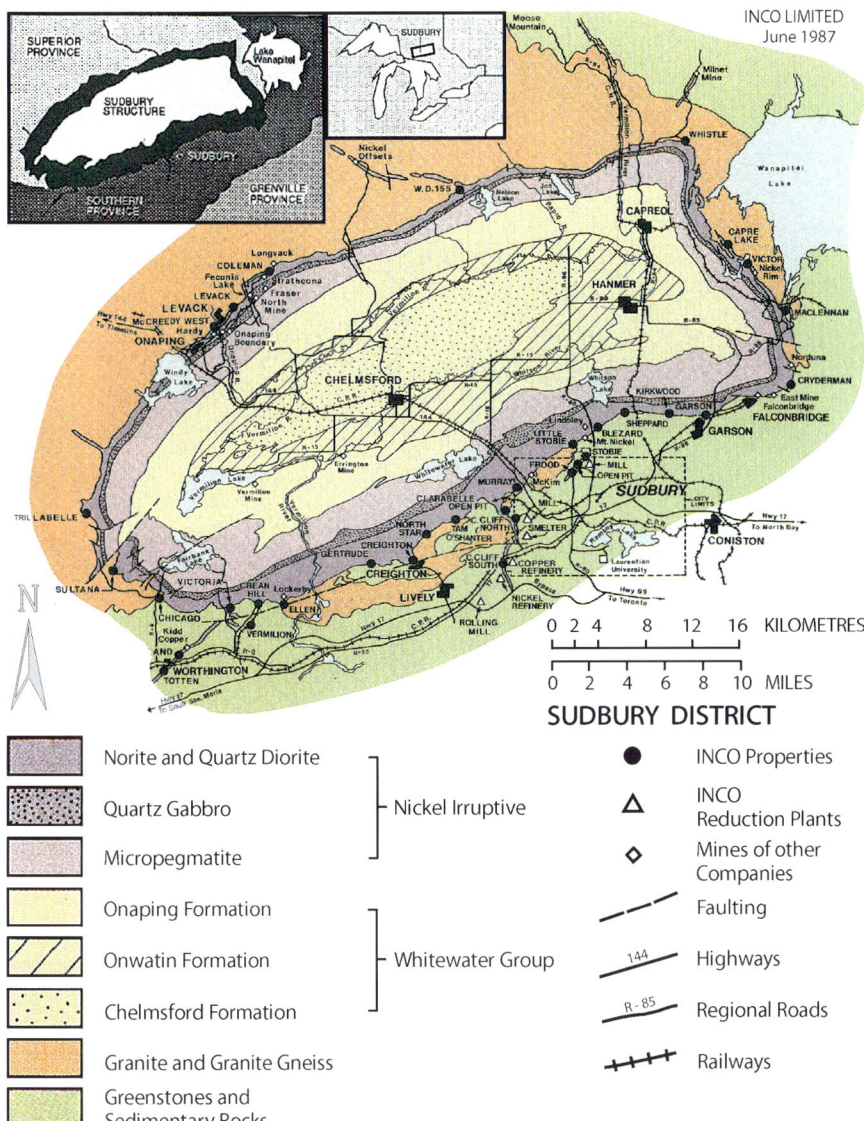

Fig. 6.34. Simplified geological map of the Sudbury Basin, Canada

landscape of today is the result of several environmental factors acting together for over a century. These factors include sulphur dioxide fumigation and airborne deposition from the mining smelters, intense logging, wildfires, water and wind erosion, and enhanced frost action (Winterhalder 1984). Mining and processing of nickel and copper started 100 years ago. As a result, soil and vegetation were lost from tens of thousand of hectares of land surrounding the smelters.

Fig. 6.35. Field photos showing vegetation damage at **a** 0 km (no trees), **b** 15 km (some dead trees), and **c** 30 km (healthy trees), downwind from Sudbury mining smelter

Figure 6.35 show the vegetation damage at the field level downwind from the smelter. It is clear that vegetation damage is extreme near the smelter and less severe at 30 km downwind from the smelter, as shown from these series of field photographs at ground level 0 km, 15 km, and 30 km downwind. Figure 6.35a shows the bare landscape near the smelter area (ground zero). This area once was covered with dense forest, but over the past 100 years, sulphur dioxide emission from the smelter destroyed nearly all the vegetation. At 15 km downwind from the smelter, the vegetation damage is not as extensive, but some of the weaker birch trees are dead from the acidic rain and soils (Fig. 6.35b). At 30 km downwind from the smelter (Fig. 6.35c), the sulphur dioxide damage obviously is less severe than at 0 km and at 15 km downwind. In addition, lakes in the region also were acidified and contaminated with metals.

The damaged area surrounding Sudbury is large enough to be detected and monitored by *Landsat TM* images. A comparison of two dates, June 1984 and 1995 were made to monitor the revegetation programs in the damaged areas. The reason for the choice of these two dates was the availability of archival *Landsat TM*

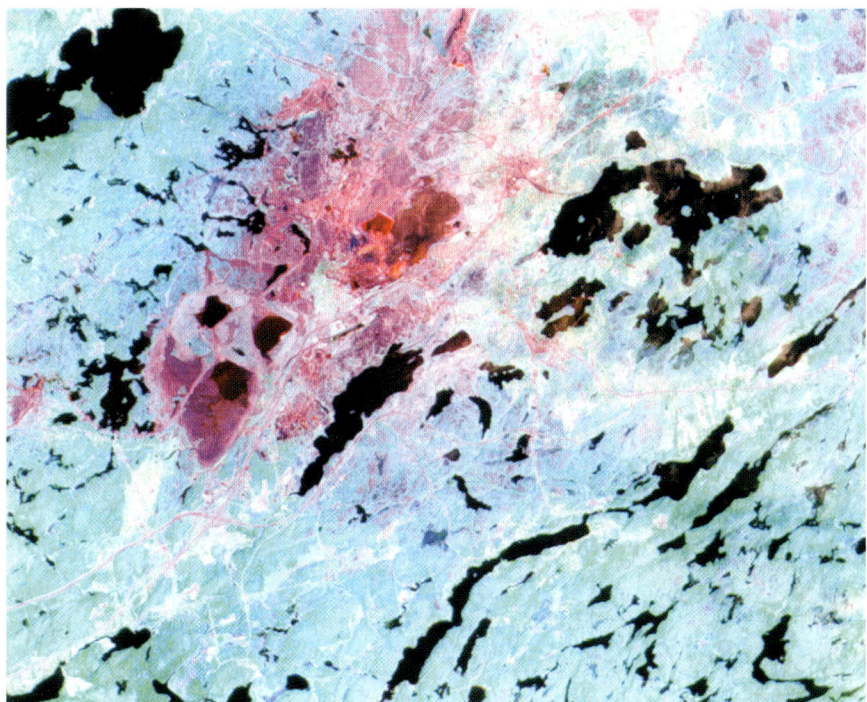

Fig. 6.36. Colour composite of 1984 *Landsat TM* bands 3, 4, and 5 (red, green, blue) of the Sudbury mining area. Area in red shows extreme damage-bare landscape, similar to field photograph at 0 km in Fig. 6.35

data over the region. It is clear to most interpreters of optical remote sensing data that the near-infrared bands are the most useful for vegetation health assessment, mainly because healthy vegetation shows bright red in this region. Using the three near-infrared bands of *Landsat TM*, bands 3, 4, 5 (red, green, blue), shown in Fig. 6.36, we were able to easily delineate the vegetation distribution and estimate the areas affected by mining, where the vegetation was destroyed. The *Landsat TM* bands were enhanced and then classified to assess vegetation change over the period 1984–1994. Standard image processing techniques such as contrast enhancement and unsupervised classification were used. The results in Fig. 6.37 (1984) and Fig. 6.38 (1995) show that the areas covered by mine tailings on the 1995 image have been reduced by 40% compared to areas covered by mine tailings in 1984. The areas in red in Fig. 6.37 and 6.38 are covered with active mine tailings. The colour legends, produced from the unsupervised classification, also are included in the figures.

The vegetation growth on the restored mine site also has increased by 50% over the mine tailings area over this period. This indicates that the revegetation program undertaken by INCO can be monitored from *Landsat TM* images. The increase in vegetation growth over the eleven year period studied is a result of a revegetation program by INCO, combined with the reduction of sulphur dioxide

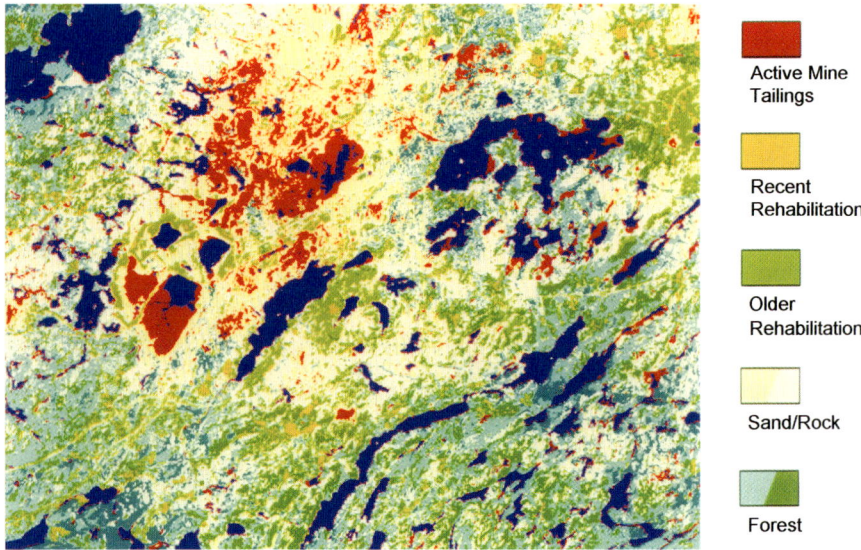

Fig. 6.37. Unsupervised land cover classification of 1984 *Landsat TM* image of the Sudbury mining area. Please refer to legend to interpret the colour scheme. Note the areas in red are active mine tailings

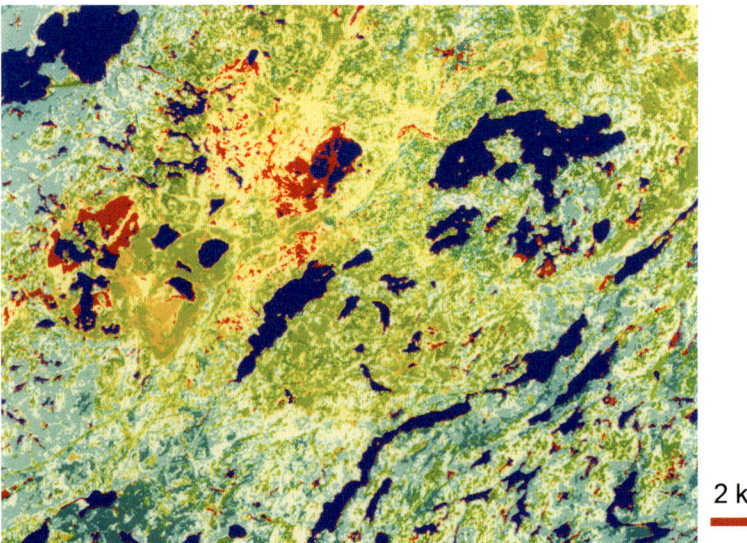

Fig. 6.38. Unsupervised land cover classification of 1995 *Landsat TM* image of the Sudbury mining area. Please refer to legend to interpret colour scheme. Note that the areas in red (active mine tailings) have been reduced, compared to the 1984 images, indicating the success of INCO revegetation programs

emission from the smelters. Since 1979, over 2 million trees and shrubs have been planted in a 3 000 ha area of limed soil occupied by grassland and open birch woodland (Beckett et al. 1995). In addition, metal levels (Ni, Cu, As) have decreased despite an increase in production at the smelter complex. These levels appear to be decreasing over time in surface soil and vegetation (Negusanti 1995). As an example of one of the many strategies, INCO now retains 90% of the sulphur from the mined ore, which is the reverse of the 1960s when 90% of the ore's sulphur was emitted into the atmosphere as sulphur dioxide (Gunn 1995).

6.5.2.2
Site Characterization and Rehabilitation of Mine Tailings

The INCO Copper Cliff tailings area is the largest repository of acid-generating tailings in North America, and possibly in the world. The site covers approximately 5 500 ha and contain more than 10% of all tailings in Canada. Mill tailings have been deposited in the Copper Cliff Tailings area since 1937, and future tailings ponds have been designed for more than 30 years of use. The impoundments contain silt- to sand-sized tailings, including sulfide minerals such as pyrrhotite, pentlandite, chalcopyrite and pyrite (McGregor et al. 1995). The mill waste is impounded in a basin enclosed by bedrock ridges and dams made of till material. The tailings dams were designed to seep in order to maintain a low phreatic head for structural stability (Puro et al. 1995). The sulfide minerals contained in the tailings are oxidized by atmospheric oxygen. A 1–3 m thick oxidation zone is present at the old tailings surface (De Vos et al. 1995).

The water infiltrating the tailings surface will ultimately seep as acid drainage. A zone of oxidized recharged water also contains high concentrations of acid drainage. The oxidized recharge water, the mill process water, and the original aquifer water are mixed and eventually seep below the tailings dam. Seepage from the toe of the dams and surface runoff is collected in seepage ponds and pumped back to the tailing ponds or channelled to the Copper Cliff Waste Water Teatment Plant. Some of the acid drainage accumulates in depressions below the dam, and gradually kills the nearby vegetation.

Airborne multispectral images (*CASI* data) were used to monitor revegetation success and to detect areas affected by acid mine drainage and seeps, ponding, surface erosion, and exposed and vegetated tailings at INCO's Copper Cliff tailings area. *CASI* is an airborne pushbroom imaging spectrometer with sensitivity in the visible and near infrared (430–870 nm) portion of the spectrum (Anger et al. 1990, Babey and Anger 1989). *CASI* operates in both a spectral and spatial mode. In the spatial mode, the sensor records up to 15 programmable bands, each having a band width of 1.8 nm. The spectral mode maintains full spectral resolution of 288 bands, encompassing the entire wavelength range of 430 to 870 nm (cf. Section 3.2.2.4).

Restoration of the tailings area involves treating the acid soils with a lime- rich fertilizer and revegetation with varieties of grasses and legumes. Old tailing surfaces consists of sparce white and yellow precipitates with mixtures of gypsum, elemental sulphur, and jarosites, which can be seen in both vegetated and non-vegetated areas.

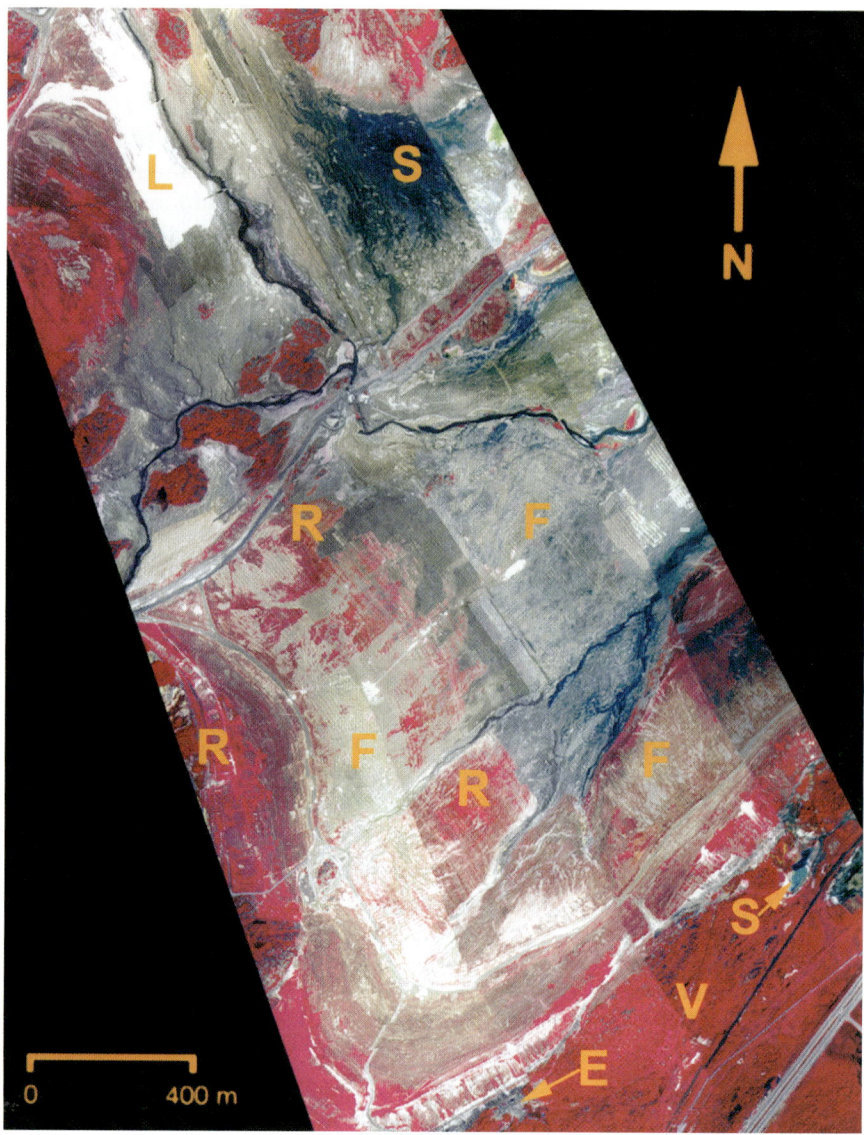

Fig. 6.39. High resolution *CASI* colour composite of bands at 680 nm, 713 nm, and 733 nm (red, green, and blue, respectively) showing areas of revegetation successes and failures over a mine tailings area (letters on image are explained in text)

The accumulation of acid mine drainage and associated stressed vegetation (S) are shown on the *CASI* composite image (Fig. 6.39). This composite *CASI* image (680 nm (red), 713 nm (green), and 733 nm (blue)), with spatial resolution of 4.2 m, is useful for identifying areas of revegetation successes (R) and failures (F) and areas of recent liming treatment (*L*). Monitoring of slope erosion and stabilization practices at the Copper Cliff tailing dams are also an integral part of environmental management of the tailings area. The dam area is capped with a layer of fine clay soil which supports a herbaceous cover. Over time, rills and gullies which developed on the slope of the dam needed to be stabilized. The *CASI* image (Fig. 6.39 at *E*) has assisted in identifying the extent of slope erosion. In addition, areas showing the success of a natural vegetative succession are shown (at *V*).

6.5.3
Summary

The Sudbury case study is a good example of the uses of both *Landsat TM* and airborne multispectral images to monitor the effects of mining and reclamation practices on tailings areas. In this study, we show that reclamation and revegetation of large areas impacted by mining and smelter effluent, and exacerbated by other natural and man-induced changes, can easily be monitored over many years using multidate *Landsat TM* data. However, site-specific restoration programs require large scale monitoring techniques. More detailed analysis and monitoring of stressed vegetation and successful revegetation can be performed using airborne multispectral and hyperspectral systems. These systems provide higher spatial and spectral resolutions to help in identifying new or ongoing problems.

6.6
Multispectral Remote Sensing to Characterize Mine Waste (Cripple Creek and Goldfield, U.S.A.)

Douglas C. Peters · Phoebe L. Hauff

6.6.1
Introduction

Abandoned mine lands have been identified as a significant problem in the United States by both government agencies and public/private environmental groups. The primary concern is the release of toxic substances and acid waters from mines, although physical hazards also are of concern. Estimates of the number and extent of mine waste sites in the United States vary widely and generally have limited supporting field surveys. Estimates of less than 100 000 sites (such as those based on U.S. federal mineral databases) and as high as 500 000 sites have been published (e.g., Lyon et al. 1993). Clearly, the extent and number of problem sites must be known more accurately if examination, evaluation, and remediation are to proceed in an efficient manner. Although coal mining has taken place in many parts of the United States, and presents considerable potential and actual hazard for acid mine drainage (e.g., Gleason and Russell 1995; Mined Land Reclamation Bureau 1982; Renton et al. 1973), this chapter will concentrate on issues and examples for abandoned metal mining districts.

Past environmental impacts of metal mines and associated wastes have been well publicized and documented (e.g., Allen et al. 1993; Azcue et al. 1995; Azcue and

Nriagu 1995; Crawford 1995; Davis and Webb 1995; Fernandez et al. 1995; Gonzalez and Ramirez 1995; Gundermann and Hutchinson 1995; Horowitz et al. 1995; Johns 1995; King 1995; Mining and Minerals Branch 1994; Rowan et al. 1995; Salomons 1995; Sengupta 1993; Stewart and Severson 1994; U.S. Forest Service 1993; Vangronsveld et al. 1995). As noted by Durkin (1995), "existing sulfide problems at abandoned mines are in need of innovative, less expensive cleanup techniques if a lasting solution" (to utilization of sulfide mineral deposits in general) "is to be achieved." The first step in making cleanup techniques cost-effective and timely is in understanding what parts of a waste site, if any, are causing problems (Struhsacker 1995). Remote sensing allows prioritization of sites so that field efforts can focus on wastes identified as the most significant environmental problems.

Waste characterization and site prioritization has become a high-interest issue within the U.S. Department of the Interior (DOI) because many of the nation's noncoal waste sites are on land managed by DOI agencies. Also, many old mine sites are on lands that are presently partially or completely managed by the U.S. Departments of Energy and Defense. Because of expertise in mining and mine-remediation technology, the former U.S. Bureau of Mines (USBM; closed by the U.S. Congress in February 1996), in cooperation with the USGS and other agencies, researched techniques for inventorying and characterizing mine wastes to allow more accurate and complete estimates concerning wastes and associated environmental problems.

Remote sensing techniques for waste characterization, involving satellite and aerial images and photographs, are particularly useful because they offer a relatively rapid method for making accurate and reproducible estimates of the number and extent of mined lands over large areas. Conversely, field updating of existing estimates, or even initial data collection, can be a long, labor-intensive, and costly process.

The objective of the research discussed in this section was to determine what remote sensing data and techniques are applicable to inventorying and characterizing abandoned mine wastes and lands. Examples of mining districts that have been evaluated through remote sensing techniques will be used to show the advantages and limitations of some techniques that were available to us for these areas. In addition, the potential impacts of the wastes on the surface and ground water hydrologic systems of the mining districts will be discussed.

The *remote sensing approach* to characterizing mined lands involves both spectral and spatial aspects of the sensors and study areas. In the research discussed here, the spectral aspects were of primary interest because of the potential geochemical impacts of mine wastes on the environment. Although the geochemical hazards are in part controlled by the physical setting of the wastes (e.g., waste volume and location with respect to drainage flow paths; e.g., see Rowan et al. 1995), mineralogy is the primary factor controlling what geochemical hazards may be present.

In this study, the *spectral characteristics* of a mining district and its associated mine sites are determined using satellite images, primarily *Landsat Thematic Mapper* (*TM*). A number of other sensors could be used, such as the *JERS* (*Japan Earth Remote Sensing*) satellite series, but the TM data were more readily available. Satellite images are preferred for the first step in characterization of mining districts and sites because of their large areal coverage and worldwide availability. This reduces the time and amount of data necessary for the initial phase of identification and characterization.

Once general satellite spectral characterization is completed, both to differentiate mined from unmined lands and determine locations of mine wastes, more detailed spectral studies of the wastes can be conducted using airborne hyperspectral sensors and ground studies. The airborne sensor used in our studies was the *AVIRIS* (Airborne Visible-Infrared Imaging Spectrometer) system flown by NASA (National Aeronautics and Space Administration). This system has 224 spectral bands (approximately 12–15 nm spectral resolution per band) over the range of about 400–2500 nm, a swath width of approximately 11 km, and a ground resolution of 20 m (Vane et al. 1993; cf. also Section 3.2.2.4). The cost of acquiring *AVIRIS* data remains significant, so these data may not be favored for general inventorying and overall characterization of broad mining regions. However, *AVIRIS* data are far superior for detailed mineralogical and geochemical characterization of selected sites. Because a single *AVIRIS* flight collects large volumes of data (11 km wide and as much as 180 km long), it often is possible to acquire information over multiple sites and distribute the costs accordingly.

Ground studies following the processing and analysis of satellite and airborne images are used to verify the mineralogical characterizations and potential geochemical hazards identified from the remote sensing data. For this study, many more waste samples were collected and analyzed per site and district than would be normal in an operational characterization program. Portable spectrometers, such as the *PIMA-II* infrared spectrometer, were used to collect detailed spectral information on the wastes to provide a verification of and connection between the mineralogical variability of the wastes and the spectral response of the satellite and airborne sensors. Depending on the type of spectrometer used, both individual waste fragments and entire waste piles can be analyzed (e.g., using the spectrometer to collect close-up spot measurements, an overall "sweep" of the waste feature by continuous collection while walking, or overall reflectance measured from some distance away from the feature with spectrometers capable of such data collection).

Chemical analysis of collected samples is necessary to verify the mineralogical and chemical associations. These chemical analyses can be performed in a laboratory through normal analytical techniques or in the field using portable X-ray fluorescence devices (e.g., see Munts et al. 1993).

In a general sense, the *spatial characterization* of mined lands is a part of the inventorying process. Any sensor which can discriminate the spectral characteristics of mine wastes from those of surrounding unmined lands also can be used for identifying the number of waste sites and (depending on the spatial resolution of the sensor) the number of individual piles and structures at each site. The spatial characterization of specific mine sites and wastes is an important factor in identifying potential physical hazards to the surrounding environs and populations and on-site hazards. Such physical hazards include slope and waste impoundment instability, unreclaimed highwalls and pits, unstable buildings and other structures, open shafts and stopes, and potential for subsidence or catastrophic collapse at a site. The issue of remote sensing of physical hazards will not be covered in this section, and the reader is directed to other sources for background information on this topic (e.g., Boldt and Scheibner 1987; Glass and Schowengerdt 1983; Mausel et al. 1981; Maxwell 1974; Schuchman et al. 1975; Zilioli et al. 1992).

6.6.2
Investigation Methodology

The *Landsat TM* data were processed to enhance surficial features, including vegetation differences, structural geology, mineral groups, and combinations of these groups. (See the introductory chapters of this book for details of the satellite systems.) Some data combinations are more useful than others, but all images provide useful characterization information when analyzed together (rather than using one isolated image).

The simplest combinations of single spectral bands, such as composites of bands 4, 3, and 2 (color-coded red, green, and blue, respectively) or 7, 4, and 2 (coded in the same order) provided information on general rock types, vegetation, and geologic structure. Color variations aid in determining geologic, cultural, and vegetation patterns. Analysis of the geologic structure is very useful in both mineral exploration and in identifying potential flow paths for ground water that could be impacted by acidic water or dissolved toxic substances (e.g., heavy metals).

Combinations (or "composites") of band ratios provide better information on mineralogy, within the spatial and spectral limits of the *Landsat TM* sensor. Band ratios involve the division of one band by another, on a pixel-by-pixel basis, and enhance specific spectral differences between materials. For example, clay minerals tend to have a low reflectance (strong absorption features) in band 7, whereas iron oxides have relatively higher reflectance in that band. The opposite usually is true for band 1. Therefore, a 1/7 band ratio tends to have high values for areas high in clay minerals and low in iron oxides and low values where iron oxides are dominant. Prior knowledge of what minerals might be present in a study area can help in constructing ratio composite images that target spectral signatures of interest. For the Cripple Creek District, which will be discussed thoroughly later in this chapter, ratio composites of 3/4-3/1-5/7 and 7/4-1/5-1/7 were found to be particularly useful for waste characterization. For the Goldfield District, a 3/1-5/4-5/7 composite was found to be the best as a stand-alone combination for general mineralogical analysis.

AVIRIS data, as already noted, have a much better spectral resolution and slightly better spatial resolution than *Landsat TM* data. Recent *AVIRIS* data can be used to map individual minerals with a reasonably high degree of confidence. Where *AVIRIS* data are available, as they were for Cripple Creek and Goldfield, *Landsat TM* spectral analysis can be completely replaced.

The *AVIRIS* data for the study area were calibrated to actual materials on the ground to be certain that local variations and conditions were taken into account. Further processing produced mineral maps both for individual minerals and combinations of minerals. The agreement between these maps and the field materials was spot checked to determine how well the classifications detected minerals for waste characterization purposes. Confidence gained from these image processing methods can be applied to other mined areas where the ground truth is not so well established.

Ground truth data were gathered using portable spectrometers, primarily the *PIMA-II* infrared spectrometer. The *PIMA-II* has an internal light source with a 5–7 nm spectral resolution and operates in the 1 300–2 500 nm range. It is manufactured in Sydney, Australia by Integrated Spectronics Pty. Ltd. The infrared spectrometer was the primary field remote sensing tool in our investigations because the minerals of greatest interest (e.g., jarosite, clays, alunogen, and alunite) for mine waste geochemical characterization can be best discriminated in the spectral range covered by the *PIMA-II*. To be certain that results were consistent over the districts and for the various wastes studied, many more spectra were collected (in excess of 1 200 spectra) than would be necessary in an operational characterization program.

To relate the characteristics of the surface coatings to the internal mineralogy of the waste rocks, the samples were also broken open and spectra of "fresh" (unweathered) surfaces were collected. In the process, the presence and relative abundance of sulfides that would be of interest from a geochemical point of view for acid-production potential were noted. The importance of the minerals in the coatings for geochemical analysis will be discussed further in Section 6.6.3.1 on the Cripple Creek District.

If iron oxides or vegetation characterization is of primary or special interest for waste site characterization, then the use of a visible-range (or visible plus infrared) spectrometer is needed. Vegetation can best be analyzed, particularly variations in chlorophyll content related to environmental stress, at wavelengths less than those measured by the *PIMA-II* instrument. Iron oxides and some other minerals also have important spectral features in the visible to very-near-infrared spectral range (less than 1 100 nm), which cannot be reached by the *PIMA-II*.

Selected samples were analyzed for elemental composition, with emphasis on metals and species such as sulfur that help determine any potential geochemical hazards. Inductively coupled plasma (ICP) was used for the general elemental analysis of whole-rock waste samples. SEM (scanning electron microscope) microprobe analyses were performed to determine the chemical compositions of coatings on the waste samples. These analyses helped confirm the minerals identified via spectral measurements.

6.6.3
Case Studies

6.6.3.1
Cripple Creek Mining District

6.6.3.1.1
Background

The Cripple Creek District is located in central Colorado in the western United States (Fig. 6.40). The district lies at the western base of Pikes Peak. The towns of Cripple Creek and Victor are present on the northwestern and southern sides of the district, respectively.

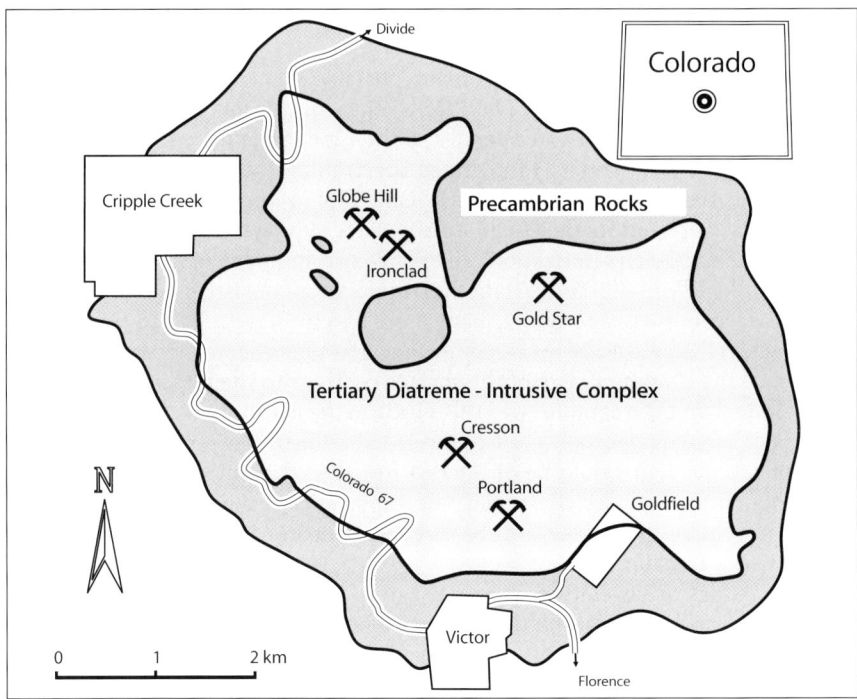

Fig. 6.40. Location of Cripple Creek mining district. Currently active (Cresson) and recent open-pit mines are identified within the volcanic complex (modified from Thompson 1986)

Geology and Mining History

The Cripple Creek District is formed by a volcanic feature of Tertiary age, generally termed a "diatreme-intrusive complex" (see Fig. 6.40; Thompson 1986). The complex also has been referred to as a "caldera" (Cox and Bagby 1986), but the current consensus appears to favor the "diatreme-intrusive" explanation. The complex exhibits occurrences of lacustrine and fluvial sedimentary rocks, subsidence features, and accretionary lapilli (Thompson 1986), thus indicating the complex was an intermittently active geologic feature for an extended period of time. Fossil leaves (Tertiary age) in lacustrine sediments have been found in mines at depths as great as 1 021 m (Koschmann 1949). Phonolite and alkali basalt intrusive rocks appear to have two different magma sources. The complex was emplaced in host igneous and metamorphic rocks of Precambrian age, and a few "islands" and ridges of these rocks were preserved within the complex. The collapse basin that forms most of the complex is divided into three subbasins (north, east, and south) separated by these islands and ridges.

The gold mineralization occurs primarily as gold tellurides, although free gold has been and can be found in oxidized and other portions of the district. The dominant alteration in the district is a quartz-sericite-pyrite (QSP) assemblage, with

Fig. 6.41. View of Ironclad Pit (in June 1994), which was mined out at this point (view from top of the waste ridge between the Ironclad and Globe Hill pits). Pit haulage road is approximately 20 m wide for scale reference

some potassic and argillic alteration. Pyrite is the dominant sulfide mineral, although it generally forms less than 2% (usually less than 1%) of the total mineral assemblage.

Low-grade, disseminated deposits formed through pressure release and brecciation of the rock as vapor-dominated mineralizing fluids moved upward in the district (Thompson et al. 1985). These deposits are currently being mined through open pits, the most recent being the Cresson Pit which began development in 1994 and is expected to be in operation for at least 15 years. The inactive Gold Star, Portland, Globe Hill, and Ironclad (Fig. 6.41) pits have associated waste dumps and leach pads that are in various stages of reclamation. The Globe Hill and Ironclad pits also have been used as waste repositories for the Cresson operation. However, at the time the remote sensing data used in this study were collected, the pits were in full operation and being expanded. During the initial stage of field work, the pits still were largely open, although Globe Hill was partly filled by waste from Ironclad and the Cresson Pit. The north subbasin of the diatreme-intrusive complex contains most of the disseminated gold breccia (such as the Globe Hill and Ironclad deposits).

High-grade veins (Fig. 6.42) and other structurally controlled deposits supported most historic underground mining (Fig. 6.43) in the district. Most of the waste dumps and mill tailings (separate from leach pads for the open pits) within the district resulted from this historic vein mining. The south subbasin of the diatreme-intrusive complex contains most of the high-grade veins in the district.

Fig. 6.42. View of No. 8 Captain Vein at the 6th level of the Portland Mine (taken in 1903) showing typical fracturing associated with the veins. Miner's candle in iron holder at center for scale. Courtesy of the U.S. Geological Survey Photographic Library in Denver and from the collection of photographs (photograph #539) by F.L. Ransome

The early gold discoveries were very rich, with grades of 30–40 oz. ton^{-1} having been common. The richest find of all, in the old Cresson Mine on the 12th level in 1914, was "The Vug". This was a cavity 12 m high by 6 m long by 4.5 m wide lined with telluride crystals and pure gold flakes (oxidized tellurides) from which the miners produced approximately 600 000 oz. of gold in one month (worth $1 200 000 at the time).

Gold mineralization was discovered in late 1890 by "Cowboy" Bob Womack, an itinerant prospector and rancher in the area. The mining district was formed in early 1891 and was a classic gold boom area.

At the time of Womack's discovery, approximately 10 people lived in the surrounding area. By 1892, more than 1 000 people had moved into the district to pursue the golden dream. By 1900, more than 60 000 people were living and working within the district and nearby related towns. Cripple Creek had approximately 35 000 residents at the peak and Victor (Fig. 6.44) had about 18 000. Cripple Creek and Victor always have been the primary centers of commerce and activity for the district. Smaller towns grew up around particular mining operations or where the boom atmosphere was not as pervasive and where "family towns" such as Cameron and Goldfield could prosper. For example, Goldfield had the first Sunday School

Fig. 6.43. View of a mined out stope (exposed during mining in the Portland Pit) in the Hidden Treasure Vein of the Portland mine showing the structure of the vein system at the top of the stope and the remaining timber supports in the stope. The stope is approximately 5 m wide. Note the yellowish and greenish jarosite coatings on the exposed rocks around the vein mineralization

Fig. 6.44. View of Victor in 1994, looking south from the Ajax mine

in the district and only one saloon in town, compared to nearly 60 saloons in nearby Victor. Cameron hosted the only zoo and amusement park in the area. The boom lasted slightly more than 14 years.

The inevitable "bust" occurred, although it happened more slowly than the boom. The World Wars were major blows to the mining industry. WWI drew many of the miners away from the area, and WWII resulted in an outright government ban on gold mining in favor of minerals needed for the war effort. In between the wars, the Great Depression occurred with its associated reduced demand for gold. An overarching factor in the demise of the old mines was the relatively steady, low price of gold which over the early half of this century did not offset the rising costs of the mining operations. Some of the biggest mines continued production into the 1950s and 1960s. However, the closure of the last gold mill in the area in the early 1960s ended all but the highest-grade underground mining operations.

Some of the old mine wastes, such as at the Cameron Leach Pad to be discussed further later in the chapter, were remined and reprocessed using a heap-leach method for recovery of the remaining gold values in the 1980s. Open-pit operations, both in the disseminated deposits and some of the vein mines, began in the late 1980s.

Cumulative gold production to the present-day is valued in excess of $7 billion based on current gold prices. Mining activity of all scales (for example, see Fig. 6.45), from prospect holes and shafts to large open pits, has been intense within

Fig. 6.45. View of upper part of Montgomery Gulch in 1903 showing high concentration of prospects and mines of all sizes and their associated wastes. Last Dollar Mine is at upper left, American Eagle Mine is on top of hill at upper left, town of Altman is on skyline at upper right, and town of Independence (abandoned in 1957) surrounds various mines at right center (courtesy of the U.S. Geological Survey Photographic Library in Denver and from the photographs (photograph # 528) by F. L. Ransome)

the approximately 35 km² district, resulting in numerous waste features, including mine dumps, leach pads, and some mill tailings.

For more information on the general history of the district, there are a number of publications available (e.g., Brown 1991; Grimstad and Drake 1983; Lee 1958). Likewise, the geology has been reported on in depth (e.g., Lindgren and Ransome 1906; Koschmann 1949; Thompson 1986; Livo 1994).

Remote Sensing Data

There were three primary remote sensing data sets available for the Cripple Creek area: a *Landsat TM* scene, an *AVIRIS* overflight image, and color-infrared aerial photographs from the NAPP (National Aerial Photography Program) of the USGS. Based on both our experience and discussions provided by Lee (1989), Livo (1994), and Taranik (1990), older *Landsat MSS* (multispectral scanner) data and *GERIS* (Geophysical Environmental Research Imaging Spectrometer) airborne data used by Livo (1994) were not revisited during this investigation because they provided little new information.

The specific *Landsat TM* scene used was #Y5 238 516 573X0 (path 33, row 33), acquired by Landsat 5 on September 11, 1990. All bands but band 6 (thermal infrared) were used in this research. Band 6 was ignored because the ground resolution was considered to be too coarse (120 m) for useful identification of mine waste features and their fine details. However, if other thermal infrared data with a finer spatial and spectral resolution were available they could prove useful in characterizing wastes in terms of their mineralogy and bulk physical properties.

The *AVIRIS* data were collected during the summer of 1989. The processing will be discussed further in the "*AVIRIS* Investigations" section later in this chapter. The *AVIRIS* data were calibrated to absolute reflectance based on laboratory spectra of representative samples from selected ground calibration sites. The signal-to-noise (S:N) ratio of the data was 30:1 to 80:1 (S:N decreased with increasing wavelength) after calibration (Livo 1994, and personal communication 1995). This relatively low S:N ratio resulted in poor data quality for mineral classification purposes. *AVIRIS* data collected today has a much greater S:N ratio and approaches that obtained by laboratory analyses. However, the 1989 data were the best available for the Cripple Creek District.

The NAPP photographs (see Fig. 6.46) coincidentally were collected on the same day as the *Landsat TM* image (September 11, 1990). This allowed them to be used as a direct check on the ability of the *Landsat TM* image to discriminate specific mining and nonmining features. The two photographs which most completely cover the district are 1063-32 (most of which is shown in Fig. 6.46) and 1063-33, and they have an original scale of 1:40 000. These photographs have a ground resolution of 1–2 m, or less for high-contrast features, which makes them very easy to use in the field (compared to the 30-m resolution of the *Landsat TM* images). The NAPP photographs mainly were used as a base for locating sample sites and a means of identifying mines and mine wastes over the entire district. The primary differences between the 1990 photographs and mine features in the district during this research (1994–1995) were the expansion of the Globe Hill and Ironclad pits (where mining finished in 1994) and the development and subsequent mining of the Cresson Pit.

Fig. 6.46. NAPP false-color infrared photograph (#1063–32) showing most of the Cripple Creek District. Note the numerous mining features (bright areas and spots) in the district and the surrounding area

6.6.3.1.2
Site Investigations

Evaluation of TM Images

As a basis for field investigations, the *Landsat TM* data were processed to produce a suite of composite images for visual interpretation and field orientation. Standard processing routines, including sun angle normalization, haze correction, and contrast enhancement, were performed on the image data. However, geometric correction was not found to be necessary because of limited apparent geometric distortions within the study area. In addition, we believed that pixel resampling

CHAPTER 6 · **Case Studies**

Fig. 6.47. *Landsat TM* false-color composite of the Pikes Peak area (at *upper right*) and the Cripple Creek District (at *bottom center*), created using bands 7, 4, and 2 (color-coded red, green, and blue, respectively). The Cripple Creek District appears as a bluish area due to the response of the altered rocks in the mineralized area (which extends to the northwest of the main district). Vegetation appears as shades of green and exposed rocks are various shades of blue and purple to white. This FCC allows some distinction between alpine tundra (such as around Pikes Peak) and mined or mineralized areas. The image area is approximately 22 km by 22 km

in the geometric correction process might unpredictably alter some of the spectral characteristics of features. Thus, we preferred not to make such corrections. Masking of vegetation was not performed because of the intermixing of vegetation and mines and waste features. The images produced were a standard false-color infrared composite (FCC; bands 4-3-2, coded as red-green-blue), a 7-4-2 FCC image, a 3/4-3/1-5/7 color ratio composite (CRC), and a 7/4-1/5-1/7 CRC.

As expected, the FCC provided little useful information on the specific mineralogy, but identified differences in vegetation, gross differences in rock types, and

Fig. 6.48. *Landsat TM* false-color composite of the Cripple Creek District, created using bands 7, 4, and 2 (color-coded red, green, and blue, respectively). The Cripple Creek District is the approximately circular feature centered in the image. The town of Cripple Creek is the grid pattern at *upper left*, and the town of Victor is the grid pattern just *right of bottom center*. Vegetation appears as shades of green and exposed rocks are various shades of blue and purple to white. Some discrimination among the various pits and mine wastes is apparent, but mineralogical differences are poorly defined with a FCC image such as this. The image area is approximately 5.5 km by 5.5 km

some alteration products. The FCCs are useful for determining field locations and providing better map bases than true-color images.

The 7-4-2 FCC (see Figs. 6.47 and 6.48) has commonly been used as an alternative to the standard 4-3-2 FCC in areas where the vegetation is of less interest than the geology. This composite was useful for field orientation, identifying differences between wastes and unmined areas, and determining gross differences among the wastes. However, delineation of the unmined from mined areas was not complete (note the similarities between the mining district and some of the al-

Chapter 6 · Case Studies

Fig. 6.49. *Landsat TM* color ratio-composite image of the Pikes Peak area (at *upper right*) and the Cripple Creek District (at *bottom center*), created using band ratios 3/4, 3/1, and 5/7 (color-coded red, green, and blue, respectively). Vegetation appears as shades of blue and exposed rocks and soils containing iron minerals are different shades of green and yellow to red. This is a typical "limonite image" produced from TM data. Note how the Pikes Peak area looks very similar to the mineralized area of the mining district because both contain considerable amounts of iron oxides in the exposed rocks. The image area is approximately 22 km by 22 km

pine tundra on Pikes Peak on Fig. 6.47). Little detailed information on the mineralogy of the wastes was derived from this composite.

The 3/4-3/1-5/7 CRC image (see Figs. 6.49 and 6.50) is the standard ratio composite for evaluating iron oxides (or limonite). The Pikes Peak Granite and other igneous and metamorphic rocks in and surrounding the mining district have significant amounts of iron oxide in their weathering products as shown in the image. As noted by Livo (1994), this composite of the mining district is not especially useful for mineralogical determination, whether for the wastes or for exploration

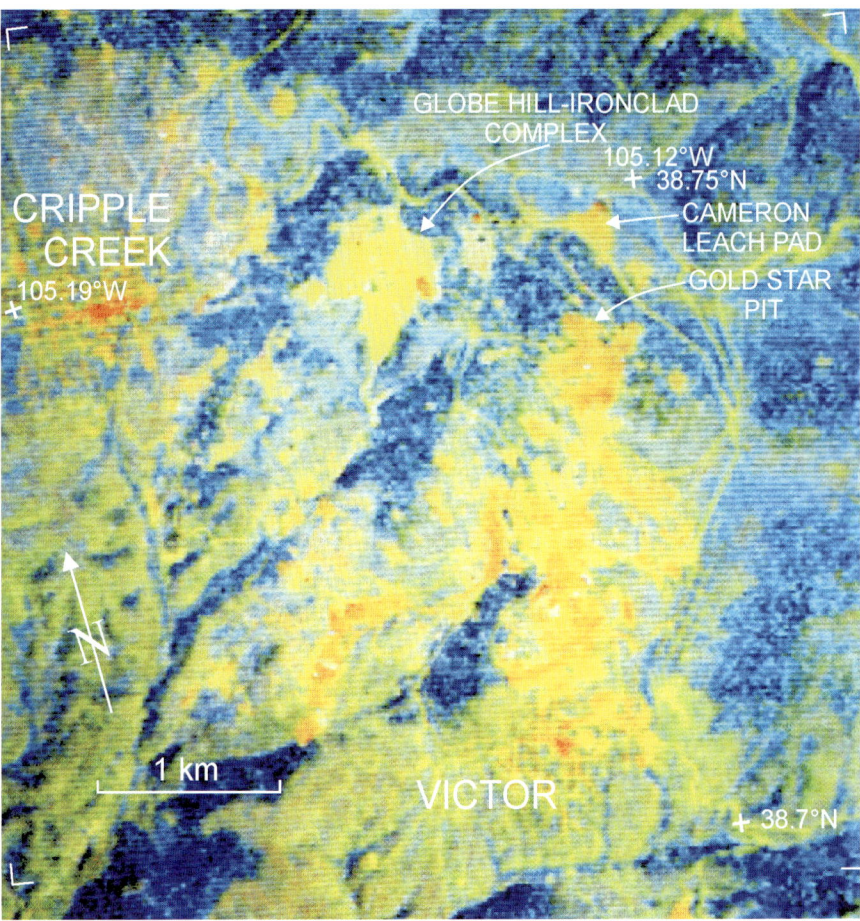

Fig. 6.50. *Landsat TM* color ratio-composite image of the Cripple Creek District and surroundings, created using band ratios 3/4, 3/1, and 5/7 (color-coded red, green, and blue, respectively). Color variations in the iron oxides (red, yellow, and green areas) are apparent; however identification of individual mineral species is limited. Vegetation appears in shades of blue (darkest where cover is thickest). The towns and northernmost open pit mines have been identified to allow the reader to match image features with Fig. 6.40. The image area is approximately 5 km by 5 km

purposes. However, it can be used as part of a suite of composites or images to provide multiple lines of evidence for the overall mineralogy of the wastes.

The synthesis of multiple image data should not be overlooked in characterization studies. The search for a single image or composite that provides all needed information in one step is likely to be futile for any study area.

The 7/4-1/5-1/7 CRC image (see Figs. 6.51 and 6.52) was devised to provide separation between the high-clay wastes and those that are dominated by iron minerals with relatively little or no clay content. It was believed that the high-clay waste, regardless

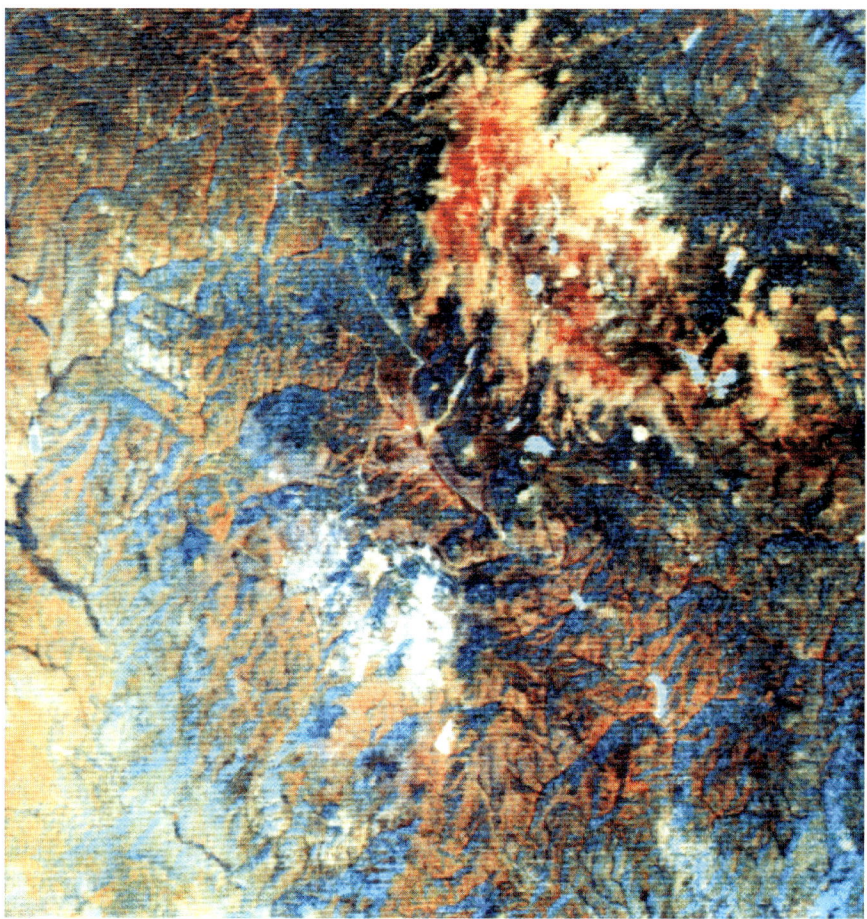

Fig. 6.51. *Landsat TM* color ratio-composite image of the Pikes Peak area (at *upper right*) and the Cripple Creek District (at *bottom center*), created using band ratios 7/4, 1/5, and 1/7 (color-coded red, green, and blue, respectively). Vegetation appears as shades of blue, exposed rocks and soils containing abundant iron minerals are various shades of yellow to red, jarosite-rich rocks appear cyan, and high-clay rocks are cyan to white. Note how the Pikes Peak area looks very different from most of the mineralized area of the mining district because Pikes Peak has little clay or jarosite in the exposed rocks. The image area is approximately 22 km by 22 km

of the amounts of iron minerals, would identify those sites most likely to contain large amounts of highly mineralized or altered rocks. Areas of most intense mineralization were believed to be the most favorable location for sulfides and heavy metals.

Altered rocks, even in many places with significant grass and soil cover, appear on this ratio-composite image as a pale blue tone compared to the more orange surrounding unaltered terrain. Where mining has occurred, the exposed mine and mill wastes and open pits appear as bright features (as do the grid-like towns of Cripple Creek and Victor).

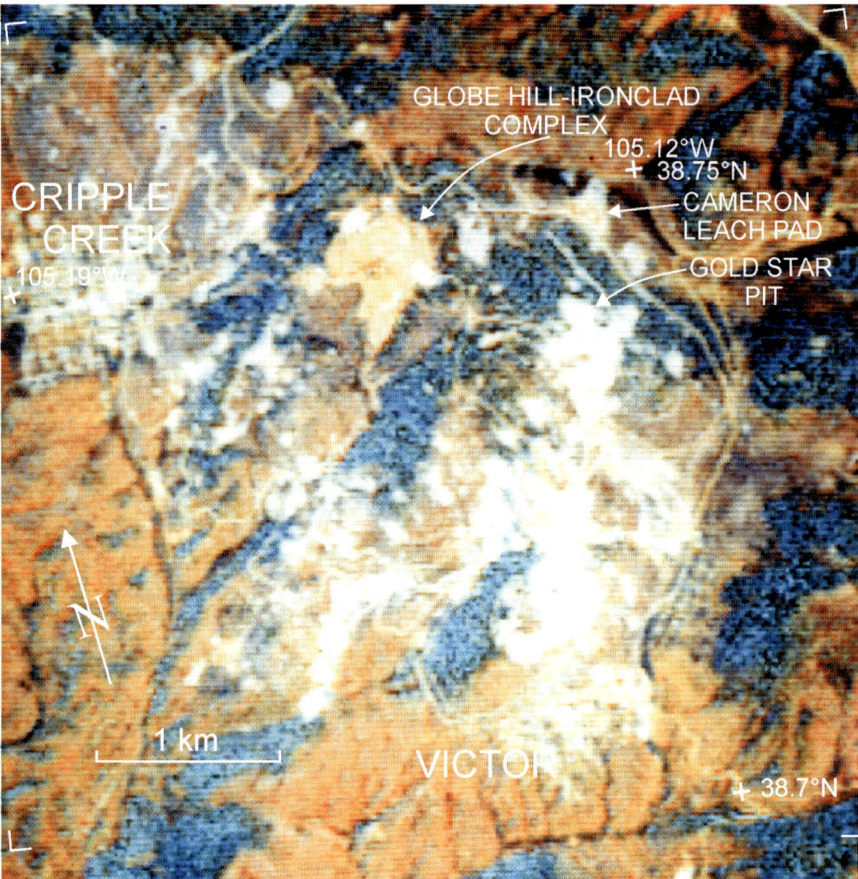

Fig. 6.52. *Landsat TM* color ratio-composite image constructed to enhance differences in clay and iron mineral contents of the wastes and surroundings, using band ratios 7/4, 1/5, and 1/7 (color-coded red, green, and blue, respectively). Red to bright-orange areas indicate low clay-content or high iron-content materials at the surface, such as weathered granite and hematitic mineralized rocks. Areas of clay alteration appear as white patches. The cyan to yellow areas indicate altered rocks that contain significant amounts of iron minerals, dominantly jarosite in the cyan areas and goethite and/or hematite with increasing yellow component in the image (based on field studies). Vegetation appears as shades of blue (darkest where cover is thickest). The towns and northernmost open-pit mines have been identified to allow the reader to match image features with Fig. 6.40. The image area is approximately 5 km by 5 km

The presence of abundant iron oxides in the wastes and open pits (both as coatings on rocks and internal minerals) appears as orange or yellow in the image. The cyan areas and general bluish background of the altered areas of the district indicate the presence of clay minerals, primarily illite with occurrences of smectite and kaolinite. The pale cyan to white areas contain a mix of clays and iron minerals, especially the iron sulfate mineral jarosite. The 7/4 ratio will have a high value where there are low quantities of clays (or where they are obscured) and sulfates and high quantities of iron oxides. The 1/5 and 1/7 ratios will tend to be relatively low under these mineralogical conditions. This results in the high proportion of red component in the image (red to yellow). Where clays dominate the spectral response of the wastes (mainly where iron oxides are low in quantity or absent), then the 1/5 and 1/7 ratios will have higher values and 7/4 will be relatively low. This results in cyan to blue colors in the image (blue increasing as iron decreases relative to clays). Where the 7/4 ratio is relatively high, indicating the presence of iron minerals, then all three ratios combine to cause a high value (pale cyan to white) in the composite image. Based on field evaluations of the *Landsat TM* images and spectrometer measurements, the dominant iron mineral in these "bright" areas of the image is jarosite. The mineralogical and geochemical implications of these responses will be discussed in the "Mineralogy and Chemistry of Waste Samples" section.

CRC ratio images provide valuable investigative information that can guide remote sensing data collection, field sampling, and other analyses. For example, knowing that the district is high in iron oxides (from the 3/4-3/1-5/7 CRC image) indicates that there is the possibility of pyrite mineralization or at least rocks that produce iron oxides upon weathering. The brighter areas compared to the vegetated background indicate surface disturbances typical of mines and wastes. The apparent response of iron sulfates in the 7/4-1/5-1/7 composite image suggest that all potentially acid-producing materials at the surface have not yet been oxidized. The presence of clays indicates locations where remaining heavy metals could reside in the wastes. Of course, details of the clay minerals which are important for understanding metal presence and mobility cannot be derived from the *Landsat TM* images. However, the *Landsat TM* image has identified areas worthy of further study by *AVIRIS* and ground methods.

AVIRIS Investigations by Livo (1994)

We have used the results of an *AVIRIS* analysis of the area by Eric Livo of the USGS (Livo 1994) to evaluate mine wastes. Therefore, this section of the chapter will refer heavily to Livo's study which used *AVIRIS* data collected in 1989.

As noted above, the *AVIRIS* data were calibrated based on laboratory spectra for selected ground calibration sites in the district. The process was as follows: "Spectra were measured for each sample on the Beckman spectrometer at the USGS spectroscopy laboratory in Denver. The laboratory spectra were averaged for each site, then a site calibration file was generated by dividing the averaged laboratory spectra by the averaged scene spectra. The site calibration files were averaged together to generate a single calibration file that was multiplied throughout the raw scene data (band for band) to calibrate the data to absolute reflectance" (p. 45 of Livo 1994).

For mineral identification, Livo used the following standard reference minerals to analyze the *AVIRIS* image and to produce fit and band-depth theme images: hematite, goethite, jarosite, kaolinite, montmorillonite (2 forms), alunite, paragonite, halloysite, illite, muscovite, gypsum, nontronite, antigorite, dolomite, chlorite (2 forms), and calcite. The standard reference spectra were derived from Clark et al. (1993).

These images were used to derive mineral identifications from the *AVIRIS* data while at the same time "considering geologic and spectral relationships and photo interpretation techniques. Each theme image was evaluated for intensity and position of anomalies." Intensities from the two types of images were used to derive mineral anomalies, where the goodness-of-fit intensity indicates how well a mineral reference spectrum matches the *AVIRIS* data and the band-depth intensity roughly corresponds to mineral abundance. Geologic feasibility of anomalies was confirmed visually (based on known geology) by Livo.

Pixel spectra from the image data were extracted and evaluated to determine mineral identification and the degree of spectral mixing of minerals. Ability to distinguish particular minerals in the data depends on the amount of overlap (or uniqueness) between absorption features of the minerals in question. The more overlap between absorption features, the less likely they will be separated in a classification process. Noise in the *AVIRIS* data can also adversely impact mineral mapping.

Some mineral mixing was found in the pixel spectra. To minimize this effect in the mineral map composites, only the highest 10% of the goodness-of-fit values were used by Livo. These materials were further enhanced in the image by applying a contrast stretch to make the top 5% brighter than the lower 5% and to make the top 1% the brightest of all. Shadows and large areas of vegetation also were masked to increase confidence in the mineral anomalies. Note that restricting the goodness-of-fit values forces mineral identification for each pixel not masked out, even if that pixel does not contain one of the end members.

Two alteration mineral image maps were produced on the basis of the minerals most abundant in the image. One image map shows the clay alteration (Fig. 6.53). The other map, the limonite distribution, is not shown here because it would not reproduce well for publication.

Most of the field verification in Livo's study focused on the open pits because of the relatively fresh exposures and the emphasis of the study on ore deposit mineralogy. Mapping limonite minerals correlated fairly well with what was found in the field, although further identification of discrete minerals (hematite, goethite, jarosite) was more problematic. Areas of "pure" goethite were clearly visible on the image. Hematite and jarosite were also identifiable. In general, spectral and physical mixing of the limonite minerals often prevented identification of the dominant mineral. The clay image map reliably identified most sericite/illite and kaolinite localities.

In general, broad mineral-association correlations exist between the thematic maps and ground mineralogy. However, the spectral and physical mixing of minerals (combined with the 20-m ground resolution of the *AVIRIS* system) and low S:N value of the *AVIRIS* data reduced the confidence in the *AVIRIS* mineral maps. The *AVIRIS* data proved more useful for mineral mapping than the *Landsat TM* data, and were generally useful for geologic and alteration mapping. The *AVIRIS* data successfully mapped argillic and QSP alteration and determined the oxidation state of the deposit. The oxidation state is important for recognizing the deposit type as well as for understanding the potential geochemical hazards of in-place rock and waste dumps (i.e., oxidized material is unlikely to contain much pyrite).

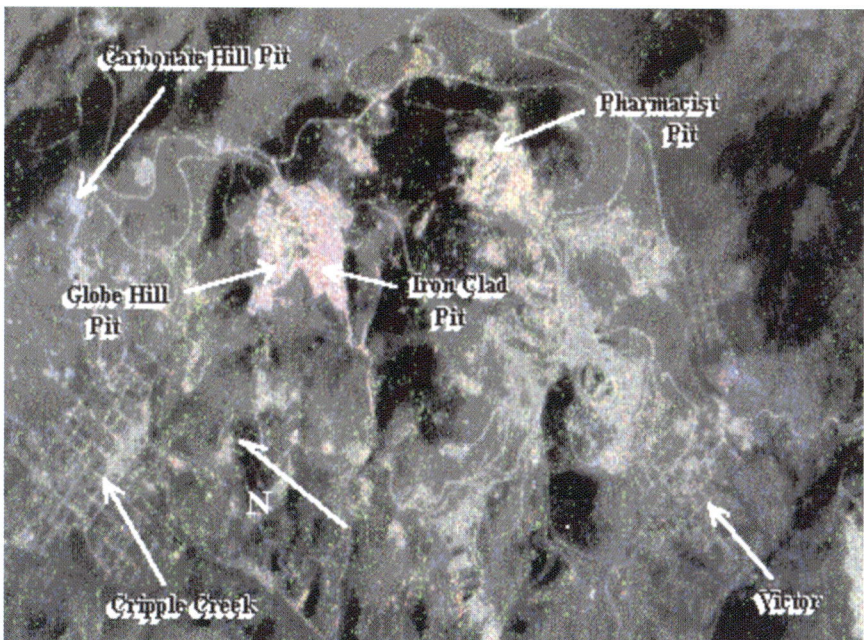

Fig. 6.53. *AVIRIS* clay alteration theme map of the Cripple Creek District (red: sericite/illite, green: smectite, blue: kaolinite). The blue and green speckle in the image is partially due to the relatively low signal-to-noise ratio (S:N = 80) of 1989 *AVIRIS* data. Significant vegetation has been masked out (black) to improve the mineral classification accuracy. Image is about 6 km by 8.8 km. From Livo (1994)

Because of major improvements in the S:N ratios of current *AVIRIS* data, it would be very useful to have the Cripple Creek area reflown. The results presented by King et al. (in this book) on mineral identifications and classifications using more recent *AVIRIS* data suggest that newer data would much more closely match what has been found on the ground.

Cameron Leach Pad Example

Evaluating the mineralogy of a waste site using *Landsat TM* data is illustrated by examining the Cameron Leach Pad (upper right part of Figs. 6.50 and 6.52). In the limonite composite image (Fig. 6.50), the site appears dominantly pale orange to yellow (very subtly mottled), which indicates the presence of limonite. On Fig. 6.52, the site has a more mottled yellow to cyan color, and is in better agreement with what was seen in the field. The chemical action of the heap leaching process, and the wetting and drying at the surface of the leach pad, resulted in most surface rocks having a dark rust-colored to black, heavy coating (spectra *F* and *H* on Fig. 6.54) of iron (Fe) and manganese (Mn) oxides which appear as yellow to orange areas on Fig. 6.50.

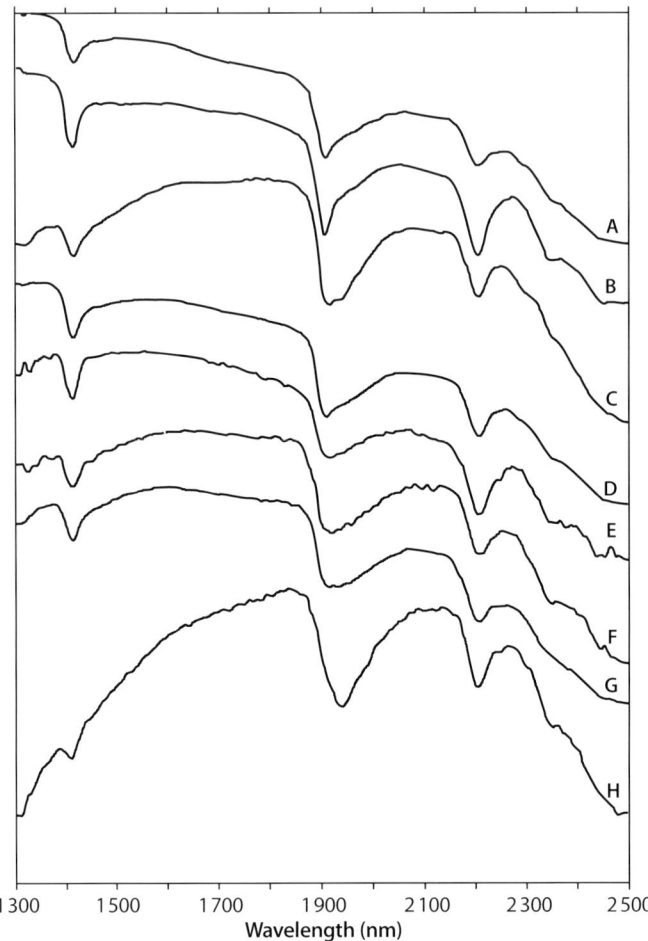

Fig. 6.54. Near-infrared reflectance spectra for samples from subsites *I* and *K* at the Cameron Leach Pad. These spectra are paired for unaltered (upper spectrum per pair) and weathered surfaces of the same rock sample. Samples *A–D* are for samples from inside the leach pad, and samples *E–H* are for samples from the surface of the pad. Spectrum *C* shows a small amount of iron oxide staining on the inside of the sample, probably reflecting complete oxidation of internal pyrite during the leaching process. Spectrum *H* is for a sample which had a heavy, rust-colored coating of iron oxides

However, trenches were dug into the surface of the leach pad during environmental investigations (Fig. 6.55; prior to the Landsat 5 data acquisition and unrelated to this project), which exposed rocks within the pad that lacked the heavy Fe- and Fe-Mn coatings. The lack of coatings indicates the possibility of different oxidizing conditions in the pad, or leaching of the oxides. Clay alteration dominates the spectra of the lighter colored internal rocks (spectra *A–D* on Fig. 6.54), and appears pale cyan to white on Fig. 6.52. The blue to cyan patch on the north-

Fig. 6.55. View of a trench in the surface of the Cameron Leach Pad. Note the dark surface material of the pad and the contrast with the light internal material of the pad. *PIMA-II* spectrometer is at bottom center. The *PIMA-II* spectrometer is field portable, has an internal light source with 5–7 nm spectral resolution and operates in the 1300–2500 nm range

east side of the Cameron site is a combination of unleached wastes, the leach-solution processing ponds, and a building.

On the *AVIRIS* clay classification image (at the very top of Fig. 6.53), the mottling results from a mixture of illite and some smectite. In the samples investigated in this research, little smectite was found at Cameron. It is possible that spectral interference from the Fe-Mn oxides, spotty jarosite on the pad, or simply mixing of clay signatures as discussed in the previous section may have occurred and confused the *AVIRIS* mineral classification process.

6.6.3.1.3
Results

Mineralogy and Chemistry of Waste Samples

PIMA-II spectrometer measurements on rocks throughout the district indicate that illite is the dominant clay alteration mineral. Depending on the reference spectrum used for mineral classification, illite often is classified as sericite (Livo 1994). Explorationists often make the assumption that sericite, instead of illite, is present in QSP alteration because the distinction between the two minerals cannot be made without detailed spectral or X-ray diffraction analysis.

Table 6.2. Wavelength of the "2.2-μm spectral feature" in samples and impact of weathering on that feature. The feature position is a function of the aluminum content (Post and Noble 1993), with higher Al at lower wavelength

Occurrence	Spectral shifts in the 2.2-μm feature by occurrence	
	2.2-μm Range	Average
Illites-pits-fresh		2.205
Globe Hill	2.202–2.205	2.205
Iron Clad	2.203–2.207	2.205
Portland	2.198–2.208	2.199
Gold Star	2.208–2.209	2.208
Cresson Prospect - weathered	2.209–2.214	2.212
Illites – dumps – fresh	2.198–2.215	2.208
Illites – dumps – fresh	2.201–2.218	2.212
Illites – weathered – poor	2.210–2.215	2.212
Illites – Mary McKinney mound	2.202–2.217	2.211
Smectites + silica – weathered	2.210–2.225	2.215
Smectites + jarosites – weathered	2.212–2.218	2.212

The illites vary in aluminum content and crystallinity with the amount of weathering (see Table 6.2). The decrease in aluminum content affects the spectra of the illites (Fig. 6.56), but this variation cannot be resolved with *Landsat TM* data and may be difficult to correlate with *AVIRIS* data.

Localized occurrences of kaolinite and smectite have been found, especially in the northwestern part of the district (Livo 1994). Smectite also occurs as a weathering product of illite where weathering has been influenced by breakdown of pyrite. Both smectite and illite have the ability to adsorb heavy metals, so knowing where these minerals are concentrated is important for understanding potential geochemical hazards.

The surface coatings of the waste samples consistently show the effects of secondary mineralization from surface exposure after mining or from contact with ground water before mining. Coatings vary from dark Fe-Mn minerals (with or without silica) to light coatings of clays and goethite. Figure 6.57 shows that Fe-Mn coatings reduce the spectral response of the clay minerals (illite in this case, based upon spectral features at 1.4, 1.91, 2.2, 2.34, and 2.44 μm), but do not obscure them entirely (see "Iron-Manganese Coatings" section for more discussion). The differences between interiors and exteriors of samples is important for understanding the chemistry of wastes.

Jarosite is the spectrally dominant iron mineral in the waste rock coatings (see Fig. 6.58; based upon spectral features at 1.47, 1.52, 1.85, and 2.27 μm) with stronger development where pyrite content is higher (based on visual examination of the

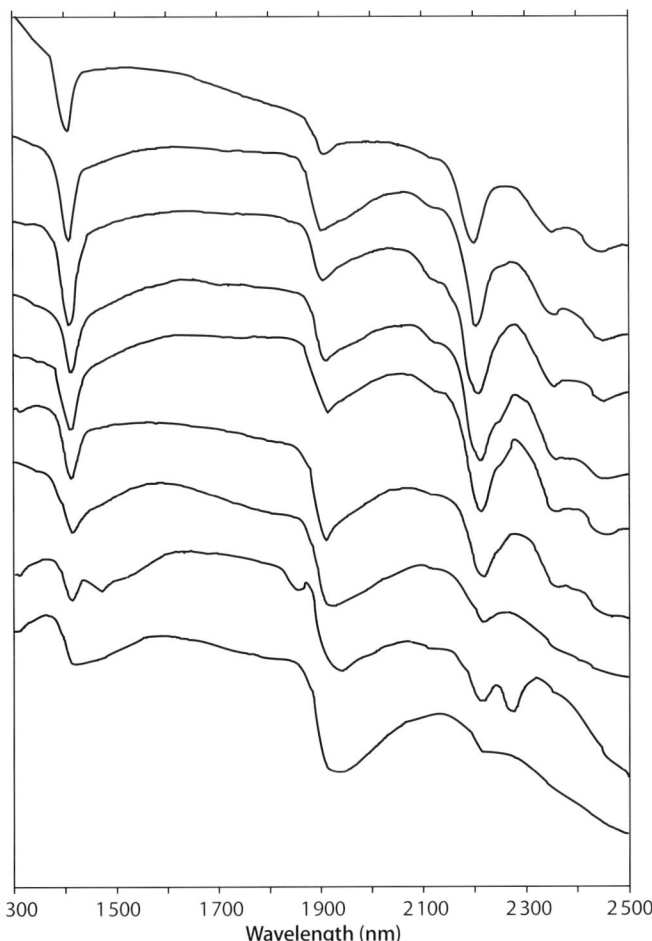

Fig. 6.56. Near-infrared reflectance spectra for various samples containing illite, illustrating how the decrease in aluminum in the illite with weathering (increasing weathering from top to bottom) changes the wavelength position of the "2.2-μm absorption feature"

samples). Pyrite content varies widely in individual waste samples, from 0% to as much as 50%. Waste samples commonly contained less than 5% pyrite and were usually in the 1% range. However, the overall pyrite content of the mineralized and altered rocks is considerably less than 2% (J. Hardaway, personal communication 1995); the high values were found in samples representing unusual vein occurrences of pyrite. The presence of jarosite was confirmed by SEM microprobe analyses (see Table 6.3).

Jarosite is a critical pathfinder mineral in determining the acid generating potential of the wastes. As shown in the geochemical series and processes in Fig. 6.59, jarosite and associated smectite require acidic conditions to form

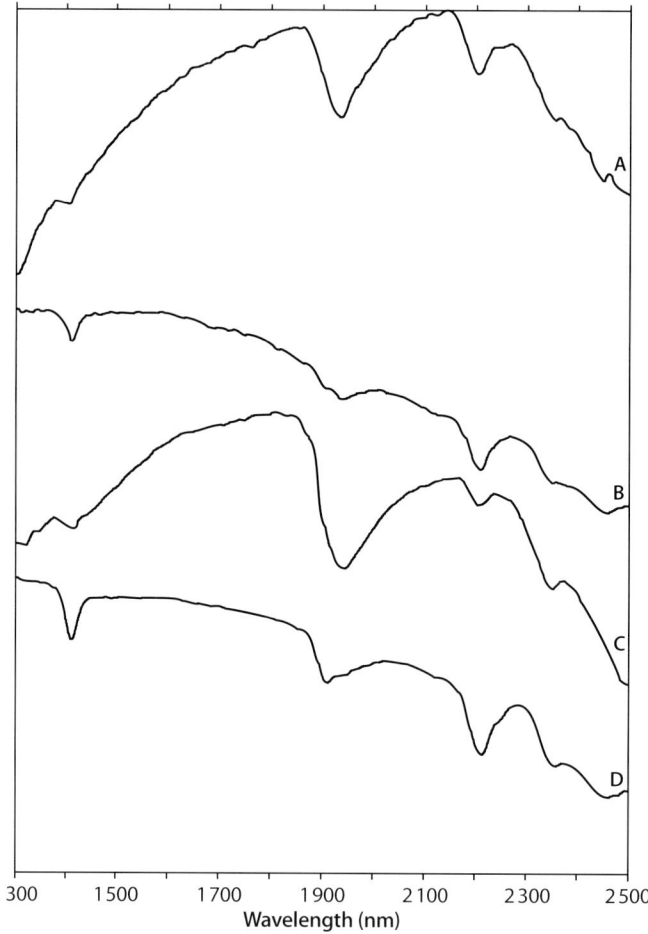

Fig. 6.57. Near-infrared reflectance spectra for two sample pairs of fresh surfaces (*upper spectrum per pair*) and weathered surfaces for samples with visually "heavy" Fe-Mn coatings. Spectra are offset vertically for clarity

from the QSP alteration and pyritic mineralization. The acidic water can be very local, formed and consumed within centimeters of the reaction site, and does not necessarily imply widespread acidic drainage conditions. Nonetheless, the presence of jarosite can be used to identify wastes that have potential for acid generation, and probably indicates that acid is being generated. Jarosite is unstable and will eventually convert to iron oxides when the acid source (mainly oxidizing pyrite and the sulfates resulting from it) is consumed or otherwise removed.

Goethite has been identified as a weathering and/or mineralization product throughout the district. Local occurrences of hematite probably are related to the

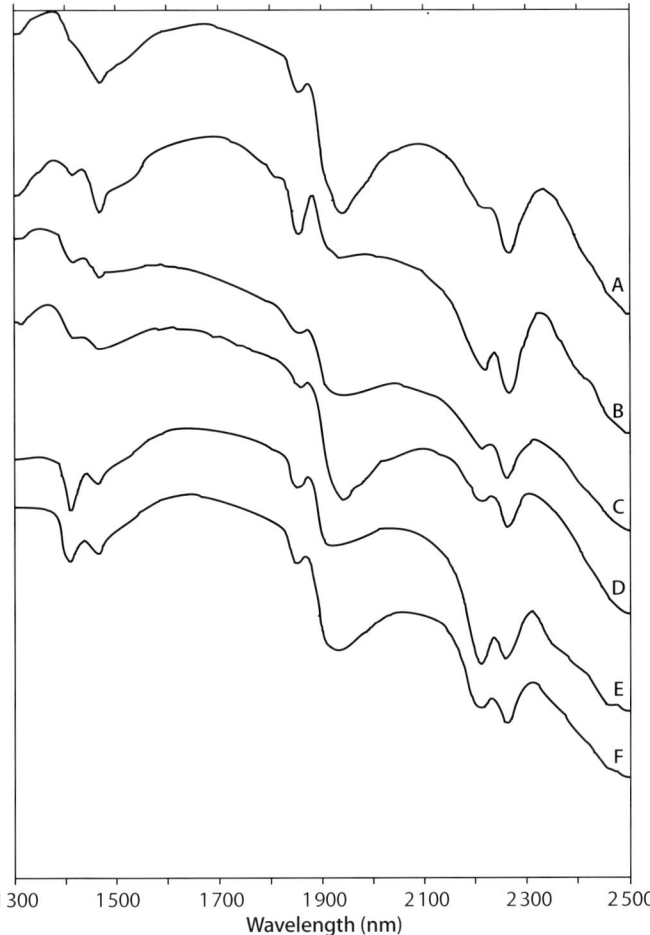

Fig. 6.58. Near-infrared reflectance spectra for typical jarosite coatings on waste samples from five sites in district. Samples *A–D* contain gypsum, *B, E,* and *F* contain smectite, and *E* contains silica. Spectra are offset vertically for clarity

original ore system (Livo 1994). Waste dumps often are composed of coarse fragments with a heavy coating of iron oxides (spectrally identified as goethite), and occasionally with Fe-Mn oxide coatings (Fig. 6.57). These areas would be expected to have little likelihood for acid production because the oxidation process likely flushed most of any mobile heavy metals from the materials.

Understanding the relation between mineral composition and the color of surface coatings can be helpful during field mapping and sampling. For example, wastes with jarosite coatings and larger amounts of internal pyrite are golden-yellow to dark green in color. Very low pyrite to completely oxidized wastes are black, red, and orange in color (see Figs. 6.60 and 6.61).

Table 6.3. SEM microprobe analyses for sulfur, iron, and potassium for jarosite coatings on waste samples

Sample	Surface Type	S (wt%)	K (wt%)	Fe (wt%)
CCWS 20-9/10	Gold-colored coating	7.81	6.06	28.56
CCCN 2	Yellow-green colored coating	7.08	9.26	28.16
CCGH 2	Yellow-green colored coating	8.22	6.99	39.13
CCGS 3	Green and orange coating	9.18	7.81	27.32
CCWS 1J	Yellow-green coating	8.60	6.81	26.31
CCWS 8A	Yellow coating	10.15	4.90	34.44
CCWS 13D	Green coating	5.61	2.80	18.26
CCWS 15B	Green coating	8.35	4.29	26.74
CCWS 15B	Orange coating	9.66	4.58	39.53
CCWS 18A	Green coating	6.36	4.93	24.22
CCWS18A	Pale yellow coating	7.87	4.70	28.26
CCWS 27A-2	Yellow-green coating	9.87	6.08	44.33

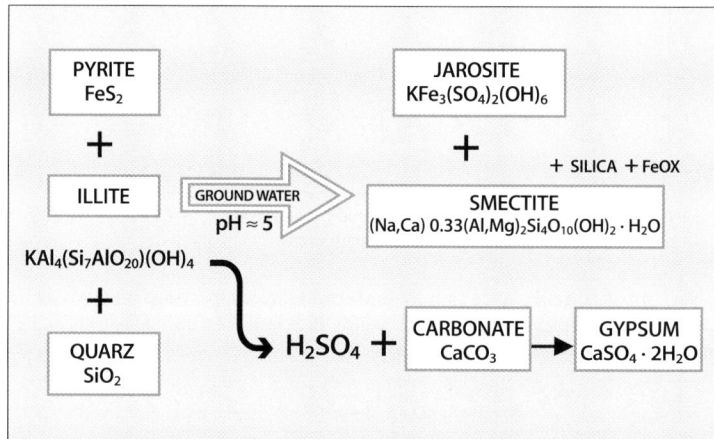

Fig. 6.59. Typical mineral weathering and geochemical series for mineralized rock containing illite and pyrite (from information in Blanchard 1968). There is loss of aluminum with weathering of illite to smectite, reflecting the measurements in Table 6.2

The analyzed samples were generally quite low in heavy metal content, (see Table 6.4). For example, arsenic exceeded 200 ppm in only a couple samples, copper exceeded 100 ppm in only one sample, cadmium consistently was 20 ppm or less, lead was higher than 300 ppm in only a few samples (most of which came

Table 6.4. Analyses (inductively coupled plasma method) of waste samples from the Cripple Creek District for selected heavy metals

Sample No.	As	Cd	Cu	Pb	Zn
CCWS 8B	124	0	134	20.8	17.1
CCWS 11A-2	218	7.84	52.2	247	509
CCWS 13C	151	0	8.31	133	48.6
CCWS 17E	102	0	20	202	241
CCWS 23-0.5	170	0	80.4	145	45.5
CCIC 1	221	8.58	17.4	346	8.06
CCIC 5	69.1	6.44	16.8	341	275
CCIC 6	181	8.38	16.1	215	0
CCIC 8	65.2	0	4.24	172	285
CCIC 14	206	14.3	54.7	115	77.3
CCIC 17	179	15.9	69.7	940	148
CCCN 1	186	3.67	7.59	24.8	0
CCCN 3	23.4	0	7.66	93.9	22.3
CCCN 5	77.6	0	0	501	0
CCGH 6	149	11	38.6	159	75.5
CCGH 7	114	15.3	41.8	958	1810
CCGH 8	271	59	50.9	654	677
CCGS 3	282	12.8	0	0	73.2
CCGS 5	225	20.7	0	0	107
CCGS 6	212	4.0	68.1	0	161
CCWS 1I	348	20.8	0	71.5	69.3
CCWS 1J	110	6.65	5.26	155	126
CCWS 1K	172	19.4	14.8	176	120
CCWS 8A	213	0	10.5	36.6	71.6
CCWS 8C	411	9.55	105	130	67.5
CCWS 12C	229	13	4.5	184	71.3
CCWS 13B	269	0	46.7	54.1	69.8
CCWS 13D	354	0	18.3	55.3	82.4
CCWS 15A	77	0	30.7	186	129
CCWS 15B	124	0	6.63	268	234
CCWS 16D	142	0	6.32	560	245
CCWS 16E	139	0	7.02	134	276
CCWS 17A	106	0	14.6	61.3	30.9
CCWS 17C	140	0	16.6	208	57.4
CCWS 18B	99.7	0	0	194	38.4
CCWS 19	166	0	12.8	235	49.3
CCWS 24B	40.9	0	5.79	59.1	5.43
CCWS 24C	59.6	0	14	31	46.8
CCWS 25	127	0	26.4	59.7	36
CCWS 27A/2	201	0	0	122	16.1
CCWS 28	123	0	14.9	73.1	76.5
CCWS 30A	217	0	6.88	177	32.1
CCWS 42C	93.7	0	0	114	199

Fig. 6.60. Typical mine wastes for the Cripple Creek District. The bright pile at left is low in iron oxides and high in clays and jarosite. The dark piles are typical waste with high iron oxides in their coatings and relatively lower clay contents and little or no internal pyrite

from the relatively fresh exposures in open-pit mines), and zinc exceeded 200 ppm in only a handful of samples scattered around the district. Gott et al. (1969) also found generally low heavy metal concentrations throughout the district. However, there still is some zonation based on structural control of the mineralization, both in general and in specific parts of the complex, on the volcanic-Precambrian contacts, and different rock types (especially between the volcanic rocks of the diatreme-intrusive complex and the surrounding Precambrian rocks).

The low metal contents of both fresh and weathered waste rocks, suggest that the likelihood for metals in the wastes or on their potential for release in any significant quantities is negligible. Plumlee et al. (1993) found very low metal contents (all <0.2 ppm) in water sampled from the Carlton Tunnel, one of the primary ground water drains in the district.

The Cripple Creek District is atypical with respect to its overall low metal content. More typical districts have significant to very high heavy-metal components to their mineralization. Emphasis should be placed on locating and characterizing wastes that have a high clay mineral content (particularly smectite) and which could provide a source for releasing metals into the hydrologic system. Kruse et al. (1989) and Munts et al. (1993) provide examples from other mining districts where this was observed.

As previously discussed, the *Landsat TM* images provide a general mineralogical characterization of the wastes. Although jarosite cannot be separated as an end member from other iron-bearing minerals or the clays in a *Landsat TM* image,

Chapter 6 · Case Studies

Fig. 6.61. Example of mine wastes in the Cripple Creek District illustrating the considerable variation in apparent colors (and associated mineralogy and chemistry) within a short distance in a given waste pile. Yellowish and greenish rocks are coated in part by jarosite related to the oxidation of pyrite in the samples

the image can be used to identify potential wastes and exposed rocks which may be sources of acid drainage (see discussion above on the 7/4-1/5-1/7 CRC). Areas with high clay concentrations and low iron oxide concentrations can be identified easily and provide pieces of the mineralogical and geochemical puzzle.

It is advisable to use the high S:N ratio *AVIRIS* data, where it is available or can be collected, to better characterize the mineralogy of mine wastes. The *AVIRIS* data used by Livo (1994) were not optimal, but could be used to identify mineralogical associations of interest both for exploration and environmental purposes. If fairly pure mineral end members occur, high-confidence classifications can be made even with such degraded *AVIRIS* data. Current high S:N ratio *AVIRIS* data can be used to map individual minerals with greater confidence, even where mixtures occur. High-confidence mineral maps from recent data could be used as a field guide to direct sample collection and chemical analyses of wastes. Such focused field work would reduce costs and the time required for site characterization and prioritization.

Iron-Manganese Coatings

An interesting sidelight of mineralogical investigations in the Cripple Creek area is the observation that coatings of Fe-Mn oxides on some wastes do not completely obscure the spectral signatures of other mineral species in those coated rocks (see

Table 6.5. SEM microprobe analyses for iron and manganese in the "heavy, black" coatings found on some wastes in the Cripple Creek District

Sample	Surface type	Mn (wt%)	Fe (wt%)	O (wt%)
CCCN 2	Dark coating	3.03	42.74	32.73
CCCN 3	Dark coating	5.02	26.40	38.10
CCGH 7	Black coating	7.35	9.62	43.37
CCGS 5	Black coating	17.10	3.68	42.12
CCIC 17	Black coating	43.98	3.91	27.41
CCWS 1K	Dark coating	19.05	1.32	31.70
CCWS 13D	Black coating	30.71	6.19	35.67
CCWS 15B	Black coating	32.55	25.09	29.22

Fig. 6.57). The coatings vary in their relative contents of iron and manganese (Table 6.5). Clay minerals in the coatings, identified with the *PIMA-II* spectrometer (Fig. 6.55), were consistent with the clays detected on fresh, broken surfaces of waste samples. The clays from the district showed weathering effects, such as decreased aluminum content in illites, consistent with the coatings having been formed through contact with oxidizing ground water or after being exposed by mining activity.

This ability to match the mineral contents of the coatings with the substrate differs from conventional remote sensing wisdom which implies that heavy Fe-Mn coatings will spectrally obscure other minerals in the rocks. For example, Lyon (1994) found that "desert varnish" from central Western Australia was composed of a mixture of iron oxide (limonite) and apparently wind-blown adhered clays (called "limonite and degraded kaolin" or LDK). This thin LDK varnish obscured the spectral characteristics of underlying rocks in many places, except where the spectral contrast with the varnish was very strong. However, Livo (personal communication 1995) reported that remote sensing investigations by USGS personnel using field portable spectrometers in the Western United States also have shown that substrate minerals can be detected through desert varnishes.

The presence of clay minerals within the varnish appears to be a common and necessary part of varnish formation. Both Dorn and Oberlander (1981) and Potter and Rossman (1979) note that illite and/or smectite clays form part of all natural "desert" varnishes. In most instances, the clays were believed, because of consistency of composition of the clays in the coatings compared to the highly variable compositions of the substrates upon which they formed, to have been deposited on the varnish-forming surfaces by water or wind. Dark Fe-Mn coatings on waste rocks in the Cripple Creek district are inconsistent with wind deposition because relatively freshly exposed waste rocks in the open pits had heavy coatings. The exposure of these rocks to wind and dust has not been sufficient to form the observed coatings. Of course, some component of wind deposition cannot be ruled out for older waste dumps or the Cameron Leach Pad. Nonetheless, inclusion of clays in the coatings through deposition by ground water (prior to mining) or by direct incorporation of substrate clays are the more likely mechanisms for the Cripple Creek area.

Dorn and Oberlander (1981) also present the intriguing possibility that most Fe-Mn oxide coatings or varnishes are formed through biological action by manganese-concentrating bacteria. The importance of this finding for our study, and remote sensing of mine wastes in general, is that such coating-forming bacteria favor near-neutral pH conditions (Dorn and Oberlander 1981). This is consistent with our expectation that such coatings form on wastes that had very little or no pyrite content as suggested by the absence of jarosite in the coatings or visible pyrite in the samples. These dark-coated (and oxidized) wastes have little or no potential for generation of acid drainage.

6.6.3.1.4
Fracturing and Water Movement

Investigations of the surface and ground water qualities and flow was not a focus of the work done in the Cripple Creek District. In general, there is relatively little surface water flow in the district, except in response to storms and snowmelt. The dominant water movement is through the ground water, no doubt reflecting the extensive fracturing systems which allowed the emplacement of the mineral deposits. What perennial streamflow exists, apart from the precipitation, appears to be maintained significantly by this ground water, both in the form of natural and man-induced stream recharge.

The primary man-induced ground water component is related to the mining development in the district. In 1910, the Roosevelt Tunnel was completed to help drain ground water from the deposits and working mines down to the 2 424-m (elevation) level. A later drainage tunnel, the Carlton Tunnel at the 2 080-m level, was completed in 1941 to facilitate deeper and drier access in the vein deposits. These tunnels drain into Cripple Creek, which is a perennial stream. Prior to the Roosevelt Tunnel, large mines were beginning to pump tremendous quantities of water up their shafts to maintain access to the deposits. For example, in 1899, the Portland #1 mine pumped 1 850 million liters of water using a large steam-powered pump at the 273-m level (depth from surface) in the mine (Grimstad and Drake 1983).

Both drainage projects helped revitalize the old vein mines. However, their positive dewatering effects were not enough to offset the negative economics of rising mining costs and the stagnant price of gold. A number of small, higher drainage and access tunnels were constructed prior to these two tunnels, but do not directly contribute to streamflow in the area.

The large drainage tunnels suggest the volume of ground water in the district and its movement. Because most of the water movement is through the rocks and their fracture systems, the neutralizing power of the host rock and mineralization can directly counteract the acid generating potential of rocks containing low concentrations of pyrite.

6.6.3.2
Goldfield Mining District

6.6.3.2.1
Introduction

Precious metal deposits hosted by acid-sulfate or high-sulfidation alteration systems have the potential to create major environmental problems because of their acid generating potential. A classic example of such a system is the Goldfield mining district in the southwestern part of Nevada (see Fig. 6.62) in the United States. Historic dumps and recent open-pit mining and heap-leach processing have created extensive exposures of acid-generating materials. These include sulfate minerals, such as alunite and jarosite, sulfide minerals, such as pyrite, famatinite, tetrahedrite-tennantite, and bismuthinite and the major mineral products of acid generation, alunogen and jarosite.

The Goldfield District is located in the western part of the Basin and Range geomorphological and structural province, 250 km northwest of Las Vegas and 300 km southeast of Reno, Nevada, U.S.A. (see Fig. 6.62). The climate is arid to

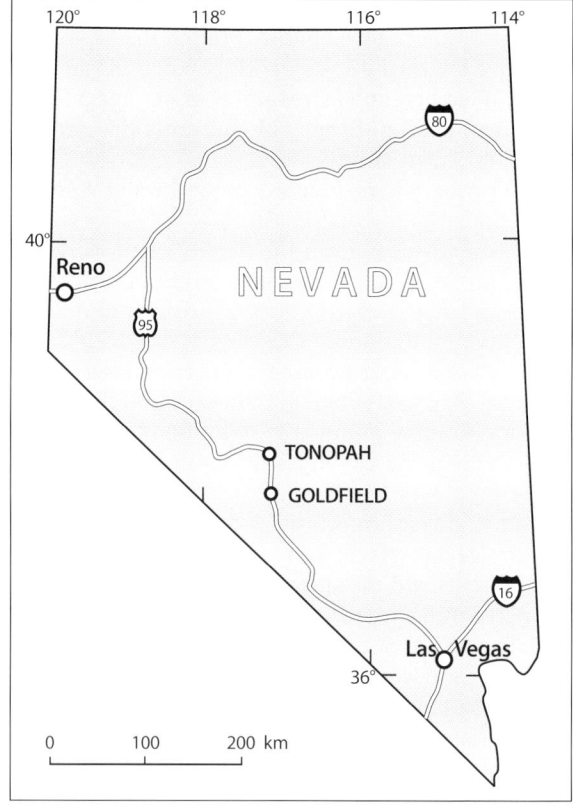

Fig. 6.62. Location map for the Goldfield Mining District in Nevada

semi-arid with limited vegetation in the form of sagebrush, rabbitbrush, various cacti, Joshua trees, and prairie grasses.

The Goldfield deposit is the largest epithermal bonanza of the enargite-gold or quartz-alunite type (Berger 1986) in the United States. Such deposits occur in calc-alkaline rocks, are associated with intense acid-sulfate hydrothermal alteration, and typically have significant byproduct copper.

At Goldfield, the ore is associated with silicified zones in early Miocene, intermediate volcanic rocks. These are termed "ledges" and lie on an arcuate fracture zone (Fig. 6.63) within extensive acid-sulfate argillized hydrothermal alteration containing quartz, alunite, kaolinite, dickite, diaspore, pyrophyllite, and illite (Ashley 1974). Ore minerals include pyrite, famatinite, tetrahedrite-tennantite, bismuthinite, free gold, and tellurides.

Gold was discovered at Goldfield in 1902. Production began the next year and peaked in 1910. The total recorded production from this world-class deposit is 4.19 million oz. (130 tonnes) of gold, 1.45 million oz. (45 tonnes) of silver, and 3 420 tonnes of copper. Open-pit mining and heap leaching during the 1980s and 1990s have yielded additional resources and in the process created extensive dumps and open exposures of the hydrothermally altered rocks. The Florence mine, shown

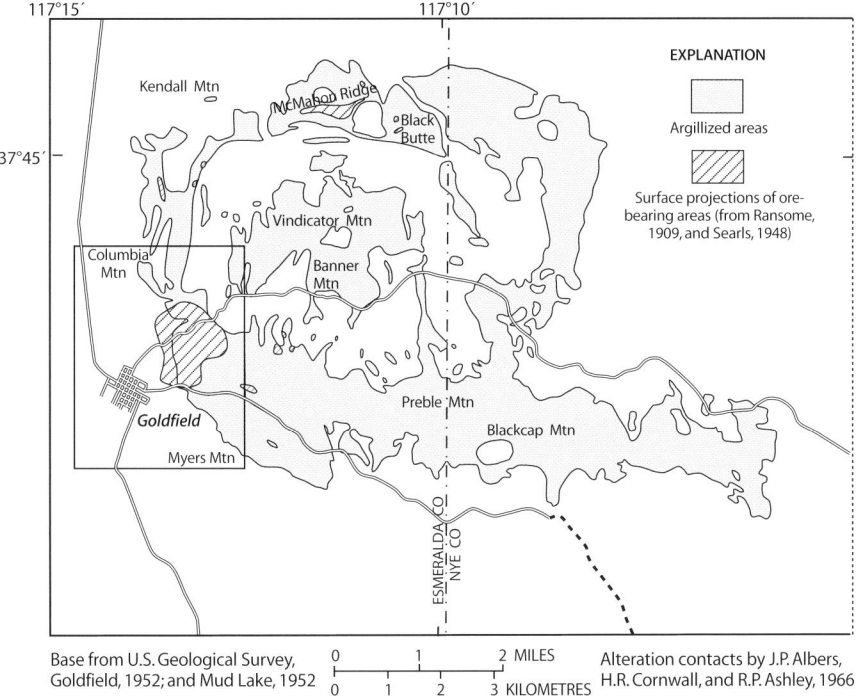

Fig. 6.63. General geographic location and hydrothermal alteration map for the Goldfield mining district (Ashley 1974). Note for discussion such sites as McMahon Ridge, Black Butte, Vindicator Mountain, Columbia Mountain, Preble Mountain, and the town of Goldfield

Fig. 6.64. The headframe and dumps at the Florence Mine, which occurs in the richest part of the Goldfield District. View is to the northwest with the town of Goldfield off the photograph to the left and the Malpai Basalt flows in the middleground. Note the arid landscape and minimal vegetation

with headframe and dumps in Fig. 6.64, lies in the heart of the richest part of the district. It still contains considerable unexploited ore. A brief background on the geology, structure, and hydrothermal alteration system will be presented to give the reader an understanding of how effective the application of remote sensing techniques has been for this mining district.

6.6.3.2.2
Geological Background

Volcanic rocks of Tertiary age host the major ore bodies at Goldfield. These rocks consist of porphyritic trachyandesite, rhyodacite, quartz latite, and rhyolite. These formed during two episodes of calc-alkalic volcanism in Oligocene and early Miocene time.

A ring fracture system created from the extrusion of the Vindicator and Morena rhyolites was emplaced in Oligocene time (Ransome 1909). Subsequent quartz latite eruptions formed poorly exposed domes along the ring fracture.

Miocene volcanic activity produced the Milltown Andesite consisting of trachyandesite and rhyodacite flows, tuffs and breccias. The Milltown subsequently was intruded by other trachyandesite and rhyodacite dome-forming events to create an extremely complex system (Ashley 1979). The Milltown was one of the host units for mineralization. These units are summarized in Table 6.6. Mineralization occurred at Goldfield within 21 to 20 my (Ashley and Silberman 1976).

Table 6.6. Selected lithologic units for the Goldfield, Nevada area (Ashley and Silberman 1976)

Rock unit	Age (mio y)	Description
Quartz Monzonite (Alaskite-Ransome)	Jurassic	Otz, Na-plug, orthoclase
Vindicator Rhyolite	33.0 ±2.0	Welded tuff
Morena Rhyolite	24.4 ±0.5	
Rhyolite flows	31.1 ±2.2	
Goldfield Latite	33.6 ±2.4	Quartz latite
Sandstorm Rhyolite	28.6 ±3.2	Airfall tuffs
Espina Breccia	22.2 ±1.4	Breccia
Milltown Andesite	21.5 ±0.5	Trachyandesite
Rhyodacite flows		
Goldfield Dacite	19.8 – 23.2	
Rhyodacite		Tuff breccia

The close temporal and spatial relations between volcanism, hydrothermal alteration, and ore deposition suggest that magmatic processes were involved with the source for the ore components found in the volcanic pile. The large amount of gold and sulfur in this system requires a deeper source.

The Goldfield District is highly faulted. The ring fracture system at one time was hypothesized to be a caldera (Ashley 1974). However, this theory has fallen out of favor. Several fracture zones apparently tangential to the ring system exist (Fig. 6.65). One, which bounds the south side of the ring fracture, trends east–southeast from the town of Goldfield through Preble Mountain. Ashley (1974) believes this fracture zone to be Oligocene in age. Another fault trends east–west along the northern extension of the district, again tangential to the ring fracture, running through the Kendall-Sandstorm cut to McMahon Ridge and Black Butte. Gold has been mined at McMahon Ridge and Black Butte, whereas a new discovery by Kennecott Mining Company was made at the western extension of this trend near its intersection with Highway 95. Just east of McMahon Ridge, this trend intersects a northwest–southeast trend which passes through Espina Hill on the east side of the ring structure.

There are myriad subsets of these fault systems, too numerous to discuss, many of which are shown in Fig. 6.65. With this extensive fracturing, the Goldfield District is highly amenable to ground water movement. This allows acidic waters to circulate in the district.

6.6.3.2.3
Hydrothermal Alteration

Hydrothermal alteration at Goldfield affected an area of more than 40 km^2 (Fig. 6.63). Goldfield is well-known for the striking ring structure configuration of the alteration systems which is apparent in Fig. 6.63. Although the alteration at

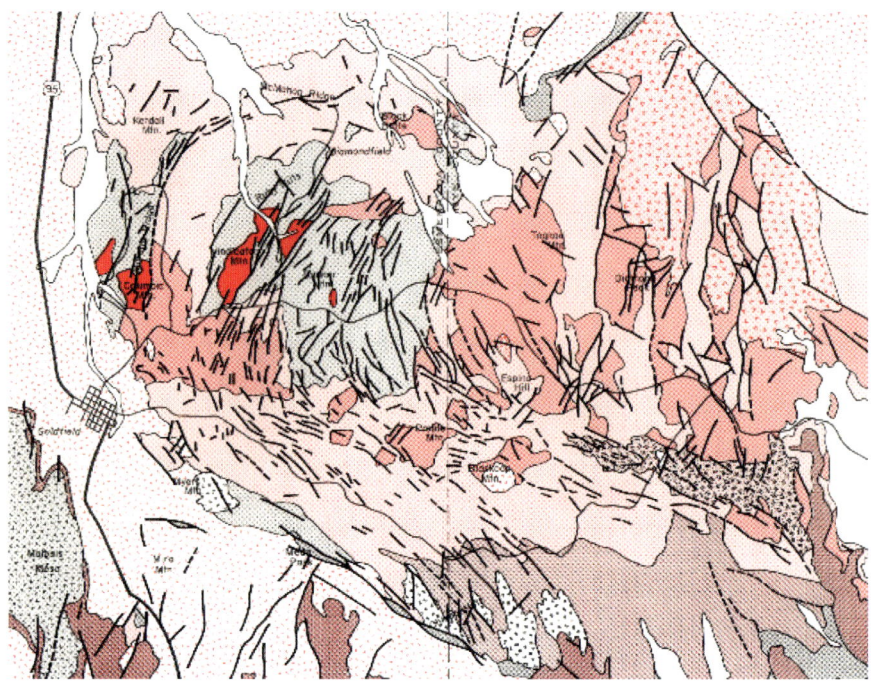

Fig. 6.65. Structural geology map of the Goldfield District (Ashley 1977). Note the three major structural trends. Most known mineralization occurs along these trends (i.e., the bright magenta and red)

Goldfield is extensive, high-grade gold mineralization appears to be concentrated in a 1.3 km² area just north and east of the town site where the majority of the production occurred. The various alteration types first identified by Ransome (1909), and further characterized by Harvey and Vitaliano (1964) and Ashley and Albers (1975), are summarized in Table 6.7.

The advanced argillic zone, as the ore host, is further divided into two subzones. The inner zone is quartz rich and is manifested at the surface by resistant outcrops. The silicification commonly follows fractures or faults to form the tabular features that are referred to as silicified ribs or ledges. This zone contains alunite and dickite along with the sulfide mineralization.

The alteration system at Goldfield has a high sulfur content and acid generating potential. This is evidenced by the presence of alunite as a major gangue mineral and the pyrite and other metallic sulfides in the ore zones.

Oxygen isotope data (Taylor 1973; Ashley and Silberman 1976), from which a δO^{18} of about –12‰ was calculated, indicate that the hydrothermal fluid was dominated strongly by meteoric water. Sulfur isotope values for the sulfides are near 0‰, which is characteristic of magmatic sulfur, thereby indicating a lower-crust or upper mantle source. The volume of gold contained in the ore deposit suggests that it could not have come from the host rocks, but requires deeper sources (Ashley 1977).

Table 6.7. Hydrothermal alteration types for the Goldfield District

Alteration type	Minerals	
Advanced argillic	Quartz ± alunite ± kaolinite ± pyrophyllite ± sericite ± diaspore ± leucoxene ± pyrite	Ore host
Phyllic-argillic	Quartz ± kaolinite ± sericite ± adularia ± opal ± pyrite	
Argillic unmineralized	Quartz + montmorillonite + illite ± kaolinite + pyrite ± relict feldspar	
Propylitic	Chlorite + albite ± epidote ± montmorillonite ± calcite ± zeolite ± pyrite	

6.6.3.2.4
Satellite, Field, and Airborne Spectroscopy Techniques

For this study, two types of remotely sensed data were used to relate the general geology with the remote sensing data: *Landsat TM* images and *AVIRIS* data. The *Landsat TM* images aided in locating sampling sites in the district. However, individual minerals usually cannot be identified using Landsat TM images. The individual minerals occurring in the hydrothermal alteration system and those created in the acid-generation process were identified with *PIMA-II* ground spectroscopy. The high-resolution *AVIRIS* sensor mapped the distribution of individual mineral species on the tailings and dumps scattered throughout this 40-km^2 area.

Landsat TM Images

A false-color composite (FCC) *Landsat TM* image was created first to provide the regional overview of the Goldfield District (see Fig. 6.66). The *Landsat TM* data were processed by Sandra Perry (Perry Remote Sensing Ltd., Denver, Colorado) using the *ENVI* image processing software package. This FCC image outlined the geologic structures and mapped the general extent of the hydrothermal alteration. The FCC was built using three edge-enhanced Landsat bands (bands 7, 4, and 1, color-coded red, green, and blue, respectively), with band-independent contrast stretching to obtain the best color balance. For the alteration study at Goldfield, band 7 is important because it responds to the clay minerals and sulfates found in acid-sulfate systems. In the FCC, high-silica exposures are shades of red, orange, and pink, vegetation is green, and water features are black to dark blue and turquoise. Bright outcrops and evaporite materials are white.

The Goldfield District is centered in this FCC image. Highway 95 bisects the image on the left. Tonopah, Nevada is off the image to the top (north). Note the red and dark brown of the Malpai Basalt flow south and west of the town and the highway. The street grid for the town of Goldfield appears in green. The arcuate structure of the alteration system is outlined in yellow, peach and salmon colors. Black Cap, which is a basalt-capped structure, shows up as a red area within the

Fig. 6.66. *Landsat TM* false-color composite image of the Goldfield District. Bands used are 7, 4, and 1 (color-coded red, green, and blue, respectively)

colored alteration. The most intensely mineralized area, directly east of the town, is flecked with bright blue. The Columbia Mill Tailings appear as a pale blue elongate feature just north of the town and east of the highway.

A second *Landsat TM* image, a color ratio composite (CRC), was produced to better enhance specific alteration patterns (see Fig. 6.67). The image consists of the band ratios 3/1, 5/4, and 5/7 (color coded red, green, and blue, respectively) with band-independent contrast stretching used to obtain the best color balance. This CRC can be used to predict general composition and alteration within the district. Iron-rich materials are shades of red and orange, acidic silicates are green, and clays are medium to pale blue. Vegetation and water features have been masked in dark blue to aid in the mineral enhancement and visual interpretation. Predicted alteration is in pale to pastel shades of pink, yellow, and blue to white. Argillic alteration trending northwest along a prominent fault zone is obvious.

PIMA-II Field Spectroscopy

Ground spectral data were collected at selected sites with the *PIMA-II* infrared spectrometer (see Fig. 6.55) to provide mineral spectra for each locality. This provide a spectral reference data set for the mineral types present. After the identification of an anomalous mineral in the *AVIRIS* image for the tailings at the Columbia Mill site, *PIMA-II* was used to field check that anomaly and subsequently identified it as being caused by the mineral alunogen.

Chapter 6 · Case Studies

Fig. 6.67. *Landsat TM* color ratio-composite image of the Goldfield District. Ratios used are 3/1, 5/4, and 5/7 (color-coded red, green, and blue, respectively)

The *PIMA-II* spectra for the hydrothermal alteration-minerals and the alunogen were imported into the image processing program(s) to further refine the Landsat and *AVIRIS* images and identify spectral components in *AVIRIS*. *PIMA-II* spectra, without atmospheric water interference and with higher spectral resolution, provide more definitive mineral identifications than the airborne instrumentation and therefore, improve the understanding of the ground information.

AVIRIS Airborne Images

The *AVIRIS* images used were processed by William Peppin (Advanced Software Applications, Inc., Reno, Nevada) using the *ENVI* software package and the SAM (spectral angle mapper) algorithm. Two images most representative of the results of this study are a mineralogical rule – decorrelation stretch image (Fig. 6.68) and a classification – best-fit image (Fig. 6.69).

The decorrelation stretch image (Fig. 6.68) shows the distribution and intensity of the hydrothermal alteration at Goldfield. It is a small subset of the Landsat image and concentrates spatially on the ring fracture. It also is rectified to a UTM (Universal Transverse Mercator) grid for ease of location of sample and district sites. For reference, note the locations (from the figure caption) for Preble Mountain, McMachon Ridge, Vindicator Mountain, and the Florence mine. Goldfield is at (UTM grid reference) North 479 000–480 000 and East 4 173 000–4 174 000.

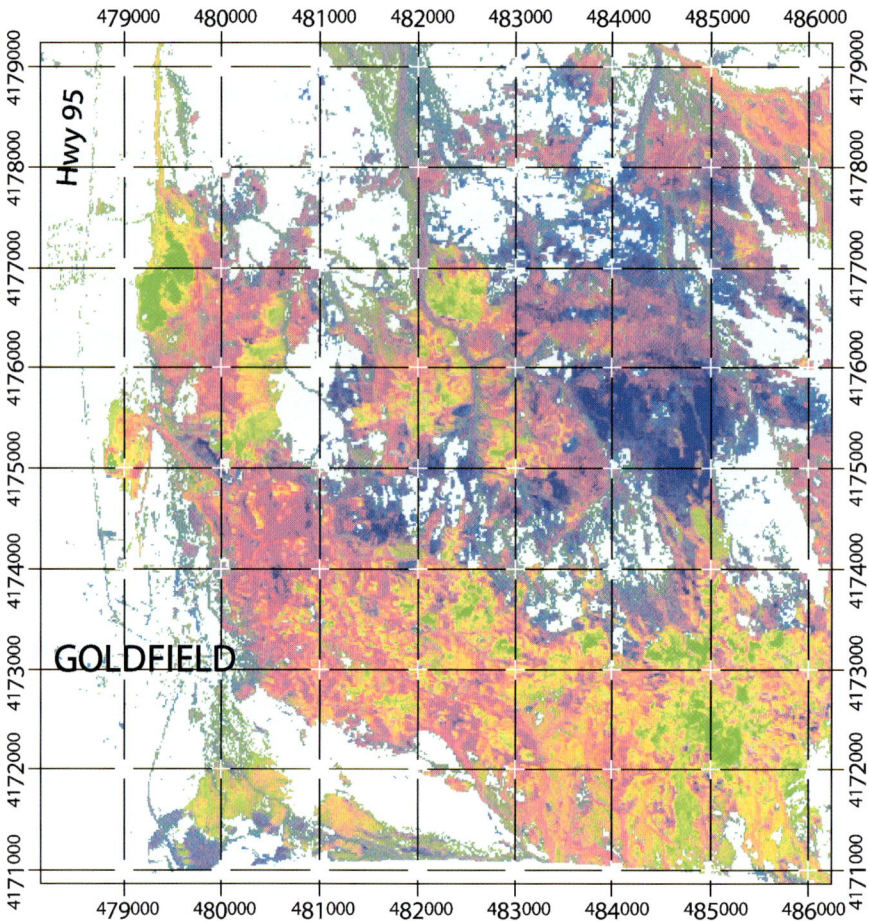

Fig. 6.68. *AVIRIS* decorrelation stretch image for the hydrothermal alteration system at Goldfield, Nevada. Minerals are color-coded as dickite: red; alunite: green; and illite: blue. Note concentration of reds and pinks in the heart of the district, just east of the town grid (North 4 174 000–4 175 000 and East 480 000–481 000). The Florence mine is located here. McMahon Ridge, with a major occurrence of dickite, is at North 4 178 252 and East 483 004. Preble Mountain, with high alunite content, is found at North 4 172 775 and East 484 962. Illite is most obvious in the center of the district, not as part of the most-altered areas

In this image, the areas of most intense alteration, indicating dickite, are shown in red and magenta. Dickite is associated with the major gold mineralization at Goldfield. Note how the red-magenta colors outline the ring fracture where most of the gold mineralization is found. Green is used to depict areas with alunite. Note the concentration of alunite on the Columbia Mill Tailings (approximately North 479 500 and East 4 177 000) and at Preble Mountain (North 4 172 775 and East 484 962). Alunite is found higher in the alteration system than dickite and can be associated with the gold, and alunite was mined historically. Thus, it appears in

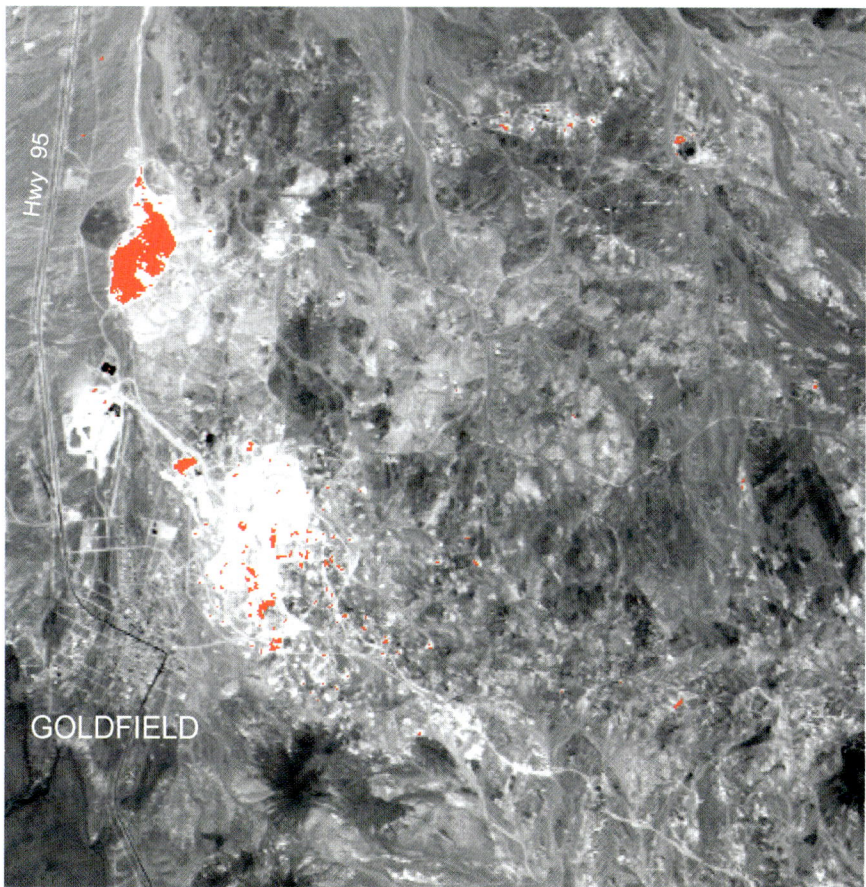

Fig. 6.69. *AVIRIS* image showing the distribution of the mineral alunogen (red). This image is produced using the 1 500 pixels that best match a library reference spectrum for the target mineral. In this case, the reference spectrum was collected on the ground at the Columbia Mill tailings site in the image, using the *PIMA-II* spectrometer

greater concentration on the older Columbia Mill Tailings rather than the newer heap leach pads (North 479 000 and East 4175 000). Although there is some alunite on the pads as seen by the yellow color, there is more of the dickite-bearing material (red to magenta). The bluer areas indicate argillic alteration, mostly illite. However, this alteration usually is not as intense. Yellow areas will contain lesser amounts of alunite mixed with argillic alteration.

The second *AVIRIS* image produced in this study is the classification – best-fit image (see Fig. 6.69). This method takes a library or reference spectrum, determines the areas within that spectrum unique to the target mineral, and then finds and fits those features within the *AVIRIS* image, creating a classified image containing only that one specified mineral.

The image in Fig. 6.69 was produced for the specialized application of determining the specific locations of the mineral alunogen. The presence of this mineral on the dumps of Goldfield was first documented using ground spectra collected with the *PIMA-II* spectrometer. The *PIMA-II* reference spectrum is plotted in Fig. 6.70 against an *AVIRIS* spectrum extracted from the image at the Columbia Mill Tailings, which is the large red area in the upper left side of Fig. 6.69. From the comparison of these spectra, it was determined that the best "unique" fits would be found in the frequency ranges of 1 500 to 1 780 nm and 2 050 to 2 400 nm. The 1 500 best pixels from *AVIRIS* data that best matched the reference spectrum were plotted as a classification image in red against a gray-scale background for the 2 100 nm band. Because the absorption features for this mineral are so broad, both spectral regions were needed for the best fit.

Alunogen is, spectrally, a difficult mineral to identity reliably because there are no specific absorption features, only a negative slope in the 1 500- to 1 900-nm region and a positive slope in the 2 040- to 2 400-nm region. Several attempts were

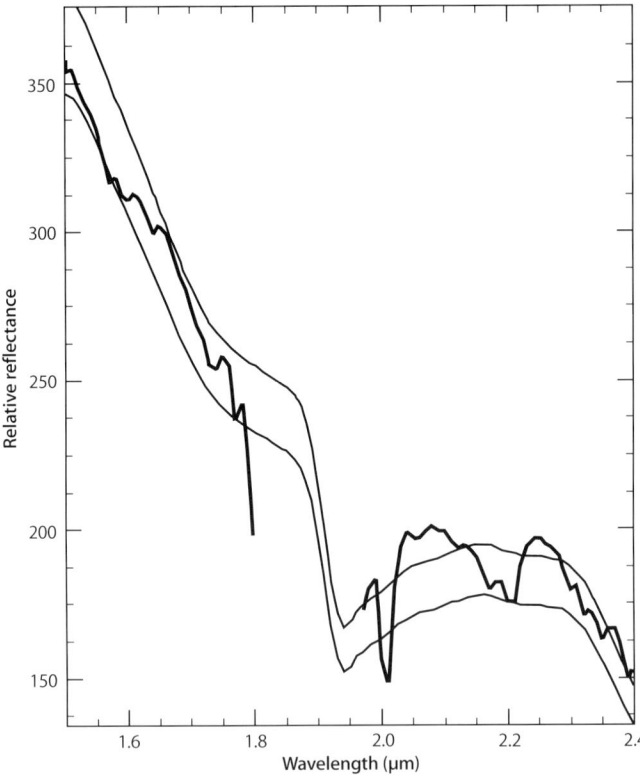

Fig. 6.70. Spectral plot of *AVIRIS* versus *PIMA-II* spectral data for alunogen. The more jagged, bold-typed curve is an extracted *AVIRIS* spectrum from the Columbia Mill tailings site which is representative for the mineral alunogen. It is compared against library spectra from the SPECMIN reference database (Hauff 1993)

made to produce a definitive image. Initial images were field checked and refined further until the authors were confident that the results were valid. The classification image for alunogen only utilized the best pixels. Alunogen was concentrated in the area of the dumps which are primary sites for major acid-generation. The exception to this can be seen for the Columbia Mill Tailings, as the large red area toward the northwest side of the image.

6.6.3.2.5
Mine Waste Evaluation and Discussion

Goldfield is an ideal area to study the potential for massive acid mine water generation. The main reason that this has not become a severe environmental problem is the dry, arid climate. The lack of high volumes of precipitation and ground water migration has reduced the volume of low-pH drainage water and, therefore, the distribution of acidic products. Goldfield still is an active mining area, with only the pits of the American Resource Company heap-leach operation currently under reclamation.

Because the acid generation has been retarded at Goldfield through lack of water, an excellent opportunity exists to track the effects of the sparse ground water activity and monitor the products developed. There is considerable "ground preparation" in the form of the ring fracture system and tangential fault systems (see Fig. 6.65), which have provided migratory channels throughout the district.

The high sulfur-bearing mineral suite including alunite, jarosite, pyrite, and heavy-metal sulfides (such as famatinite, enargite, tetrahedrite-tennantite, and bismuthinite) is the raw material for acid production. The primary sulfate mineral at Goldfield is alunite. Comparison between the Goldfield *AVIRIS* alunite image (Fig. 6.68) and the alteration map (Ashley 1974) shows the distribution of alunite. It is present in major concentrations (green and yellow on the image) throughout the alteration system. Alunite occurs primarily as soft, fine-grained, argillized replacements of hydrothermally altered feldspar phenocrysts. This makes it easy to break down through ground water processes. Alunite is associated with the sulfide ore areas and, therefore, a mineral map of alunite will indirectly reflect the sulfide distribution as well.

Mine Waste Minerals

Minerals produced from the acid-generation processes at Goldfield include the jarosite and smectite pairs seen at Cripple Creek and minerals previously discussed in Section 6.6. However, jarosite and smectite are not in the same abundance as at Cripple Creek because illite is not as common at Goldfield.

The main weathering product of acid water observed on the dumps at Goldfield is alunogen, $Al_2(SO_4)_3 \cdot 16H_2O$. Alunogen was first recognized, in an *AVIRIS* image as an anomaly at the old Columbia Mill Tailings north of the town of Goldfield, just off Highway 95, by Richard Bedell of Homestake Mining (see Site *G* in Fig. 6.71). The anomaly was field checked with the *PIMA-II* spectrometer, and the mineral was identified as alunogen. As discussed in the previous section, using this dump spectrum as a training reference, subsequent images were processed to specifically identify alunogen (Fig. 6.70).

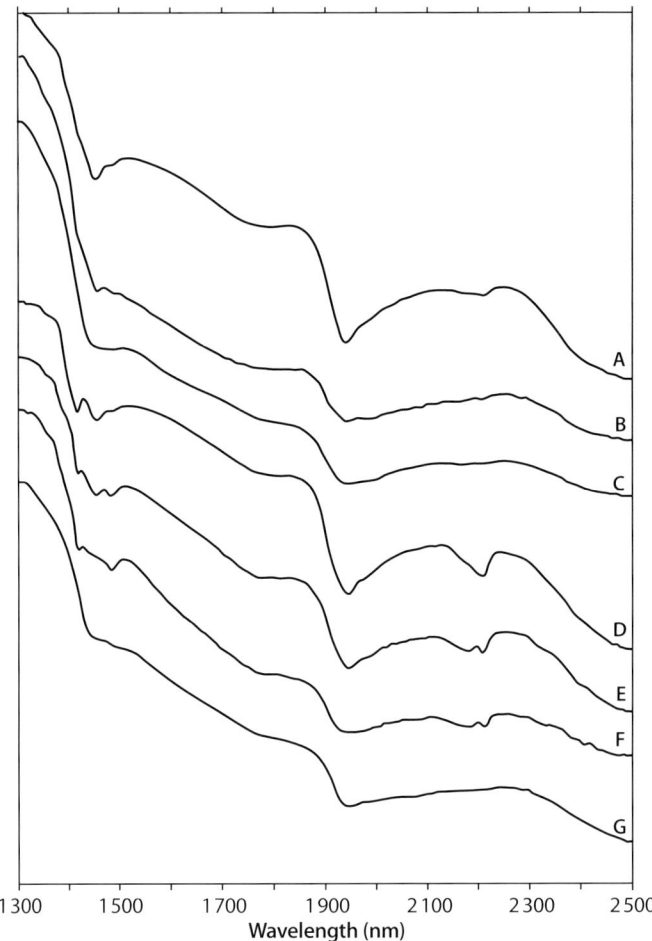

Fig. 6.71. *PIMA-II* spectral plots for various dump and tailings sites at Goldfield correlated with the *AVIRIS* image. Spectrum A is from the Red Lion dump in the southeastern part of the district. Spectrum B is from the large dump on the northwest side of Black Butte. C is from a dump just south of the Florence mine. D was found at the Deep Mines Dump in the center of the district. E comes from a large dump at the northern intersection of the two faults. F is from the Columbia Mountain Mill Tailings and contains alunite. G is from the Pittsburgh Dump

Spectra of alunogen from the checked dumps are shown in Fig. 6.71. Although all the samples contain alunogen, most contain other mineral species. This compounds the difficulty in mineral identification. Fortunately, alunogen does dominate the spectra. The mineral grows on the dump surface and is a direct product of acid weathering. The underlying minerals are not always completely covered or altered by the weathering process. Spectrum A is from the Red Lion dump in the southeastern part of the district. Spectrum B is from the large dump on the northwest side of Black Butte and contains very minor dickite. C also is from a dump in the southwestern part of the district, just south of the Florence mine. *D* was found at the Deep Mines Dump in the center of the district. It is soft, buff-colored, and contains clay, probably smectite. *E* comes from a large dump at the northern intersection of the two faults and has minor alunite. *F* was the first spectrum collected in this study, is from the Columbia Mountain Mill Tailings, and contains alunite. *G* is from the Pittsburgh Dump, is pale yellow, frothy, and siliceous, and is the purest alunogen among the sample set.

The alunogen appears in several habits on the dump surfaces. It can be frothy, siliceous, and white to pale yellow, and as such, usually is associated with a ground water deposited silica. Alunogen can be earthy and buff-colored, very soft, and associated with clays, usually a "popcorn" (referring to the ground surface texture) smectite. Alunogen also can have a "Swiss cheese" type of texture which is siliceous with large holes, but not frothy.

The association between the acid generator, alunite (and pyrite), with the weathering product, alunogen, can be seen by comparing the two *AVIRIS* images. Alunogen is found throughout the district on the older dumps in the gold producing areas, but it was not observed on the new heap leach pads, nor is it seen on disturbed dumps which are being reworked.

The type of mineral map produced in Fig. 6.69 can be used to locate the major acid generators in a district. The ability to map a mineral as difficult to detect as alunogen demonstrates that remote sensing techniques have become a very powerful and effective tool for discrimination of mine waste products.

6.6.3.3
Miscellaneous Other Sites

Similar relationships between mineralogy of surface coatings and internal mineralogy have been observed in samples from other metal mines and mining districts in the Western United States. Samples were obtained from mine wastes in the Butte District and the Crystal Mine in central Montana, from the Midnite uranium mine in northeastern Washington, from the Leadville mining district in central Colorado, from the area of the Getchell Mine in northern Nevada, and from the Summitville Mine in southwestern Colorado. The Midnite Mine is one of only two large, hardrock uranium mines in the United States, and pyrite is one of the dominant accessory minerals in the deposit. The other locations all are sulfidic precious and/or base metal deposits.

Figures 6.58 and 6.72 show the spectra of the weathered surfaces of pyritic samples. They are dominated by sulfates, illite, and smectites, in keeping with the weathering mineralogy and chemistry seen in Fig. 6.59 and discussed earlier in this chapter. Figure 6.73 shows wastes in the Leadville District (a high-sulfidation mineralization system) which have similar surface coloration and coatings common to sulfidic wastes. The generation of jarosite, in particular on sulfidic substrates, appears to be the norm in mineral districts which contain sulfide minerals.

The basic spectral characteristics observed here have also been seen in other parts of the world during activities by Spectral International Inc. Such relations between weathering products and internal mineralogy have been observed in Argentina, Canada, Chile, and Mexico and are likely to be observable elsewhere where sulfides are significant parts of the mineralization.

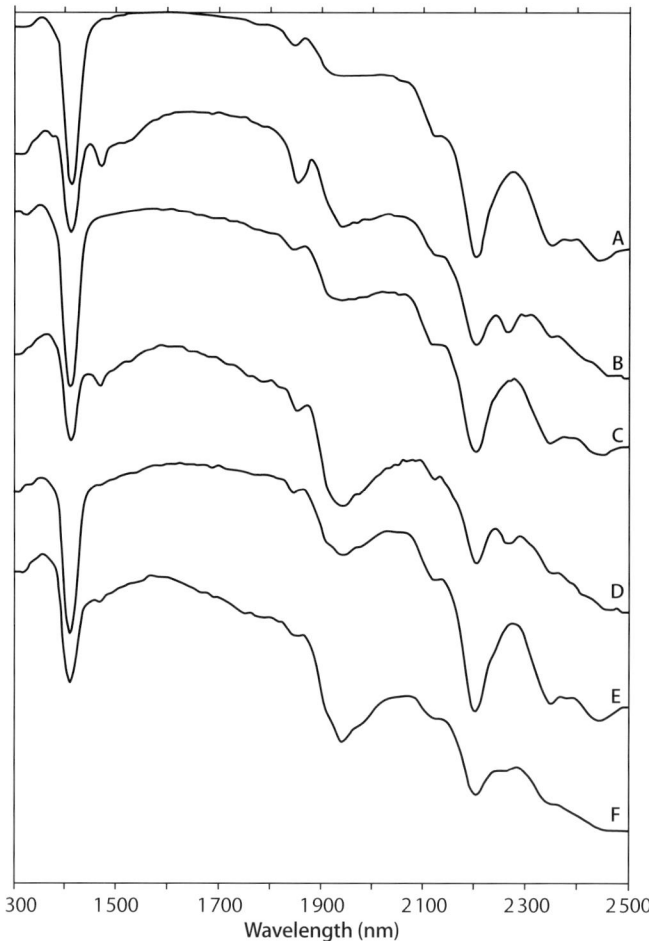

Fig. 6.72. "Weathering" pairs for samples from the Butte District in central Montana. The upper spectrum in each pair is for a fresh surface and the lower spectrum is for a weathered surface. The coatings display presence of jarosite and all three samples contained significant pyrite

Because most metallic mineral deposits have some degree of structural control (Cox and Singer 1986), the potential for migration of metals and acidic waters along structures exists. Neutralization potential of the host rock or accessory minerals in a particular deposit, and the wastes derived from it, may limit the environmental impact of surface waters and ground water seepage that have been in contact with sulfidic and metal-bearing wastes. Consequently, the presence of fracture systems and faults that intersect sulfidic or otherwise mineralogically unstable wastes should be identified through remote sensing data. Waste site prioritization should consider such ancillary information when available to fully identify potential hazard and degree of risk of abandoned mine lands.

Fig. 6.73. View of waste dump at the Moyer Mine in the Leadville District showing the typical multicolored nature of the surface of the wastes. In this pile, very-high pyrite content material is medium gray, altered carbonates are black to dark gray, and jarositic material is yellow to green in color. Oxidized material (at least at the surface) is orange to rust-red or brown in color

6.6.4
Prioritizing Waste Site Investigations Based on Remote Sensing

6.6.4.1
Data Source

On the basis of overall spectral response, mineralogy can be differentiated using remote sensing data and waste features and disturbances in mineralized areas can be distinguished from nonmining features such as paved roads, most buildings, and alpine tundra. Satellite and airborne images of mineralized and unmineralized areas have been used for mineral exploration purposes for at least the past two decades. Unreclaimed mine and mill wastes commonly are very distinct areas on images and photographs because of their lack of vegetation and incongruous topography.

Primary considerations in determining methods for site characterization are the availability and cost of data versus the information gained from a given technique. *Landsat TM* data allows a large area (even nationwide) to be inventoried rapidly and at relatively low cost per square kilometer. However, the degree of mineralogical characterization is limited. Also, small mines and prospect pits will be largely undetected or indistinguishable from their surroundings.

SPOT satellite data (see Chapter 3 on types of satellite remote sensing data for discussion of the *SPOT* system) provide better spatial detail, but are lacking in the spectral range and resolution necessary for mineralogical analysis. For characterization purposes, a compromise can be obtained through digitally merging *Landsat TM* and *SPOT* images to provide images with better spectral and spatial features than either of the individual data sets.

AVIRIS images can provide the detailed mineralogical characterization of wastes and better spatial detail. However, the higher cost of data acquisition and processing will preclude *AVIRIS* from being used as a reconnaissance and inventorying tool. The higher spectral resolution *AVIRIS* data would be useful for pre-field characterization. Imaging spectrometer satellites, due to be launched within the next few years should provide worldwide data coverage at useful spatial and spectral resolutions for waste characterization and be available at reasonable costs.

6.6.4.2
Mineralogy Versus Priority

The correlation between sulfide content and mineralogy of the wastes in Cripple Creek and Goldfield shows promise for being able to qualitatively characterize acid potential of wastes with remote sensing data. Similarly, if metal content and species of the wastes show a correlation to clay and/or iron mineralogy, the potential for heavy metal release may be identifiable through the remote sensing data.

Image data can be used to produce automated waste classifications and waste location maps. Supervised and unsupervised classification schemes are standard means of breaking image data into spectral groups. The relatively new, experimental method of artificial neural networks shows promise as a means of achieving automated analysis.

Prioritization of further investigations of mining wastes should be directly tied to the remote sensing results. The remote sensing data will allow regional characterization, and provide a means of prioritization for supplemental field evaluations on sampling. Exhaustive sampling and field mapping is very slow and prohibitively expensive considering the urgent need for nationwide waste characterizations and the limited funds available for the task.

6.6.4.3
Application of GIS Technology

Application of geographic information systems (GIS) technology to mine inventorying and characterization could improve the accuracy and completeness of image classifications. GIS analysis could relate classified features to known sites and other historical and analytical information pertaining to wastes and mining. For example, to distinguish mined from unmined areas in images, areas that appear as disturbed ground can be compared with spatial databases of cultural features such as digitized topographic maps, to eliminate unpaved roads and other known non-waste features from further consideration. GIS analysis would allow wastes to be spatially analyzed in terms of drainage basins, potential interactions with surficial and subsurface geology, proximity to populations, and other cultural, aesthetic, or environmental factors.

GIS databases provide a means of standardization of data for a given area. For prioritization purposes, such data would include analyses of site-specific liquid and solid wastes, regional, local, and site-specific map information on the geology, history, biology, and other overall conditions may impact or be impacted by mining activities. Historical and recent photographs should be considered. Such databases can be used for prioritization and remediation and serve as a basis of comparison for past and future activities.

The potential uses of GIS technology are too numerous to cover here and a complete discussion of the technology is beyond the scope of this book. The following references are recommended for a better and varied background on environmental applications of GISs: Douglas (1995); Goodchild et al. (1996); Johnson et al. (1992); and Lyon and McCarthy (1995). However, the lack of GIS application for mined lands analysis among these publications indicates the limited use of GIS (and remote sensing) technology within the mining industry, particularly with respect to environmental analysis. The applications of GIS technology within the mining and environmental industries are expected to grow considerably in the future.

6.6.4.4
Site Prioritization

Once waste features have been inventoried and characterized, the sites can be prioritized. The criteria for prioritization can include the introduction of other information, such as known mining history or sample analyses, contained in public-domain databases and in technical literature. Inventorying should not be attempted from these available sources alone because of the embedded inconsistencies in data quality. Prioritization involves determination of what sites or districts need additional analysis because of their potential or known impacts on people and the environment. After prioritization, actual fieldwork on high-priority sites should include on-site inspection and sampling for detailed chemical and physical characterization. The remote sensing data would serve as a guide for these intensive and costlier efforts.

Well-planned field sampling will maximize resources and lead to better reclamation or remediation designs. Prioritization would begin with an objective characterization of large areas and finish with analyses of specific hazardous wastes rather than treating all mine wastes equally. Failure to adequately and quickly prioritize sites leads to overly lengthy and financially wasteful cleanup projects.

6.6.5
Summary

The utility of satellite, airborne, and ground remote sensing data has been demonstrated for identifying and characterizing mine wastes. Knowledge of the mineralogy of the wastes is a critical part of the information needed to fully understand the potential and active geochemical hazards posed by wastes at abandoned and inactive mine and mill sites. Advances in airborne and portable remote sensing instruments (and, soon, satellite systems) have led to the ability to map mineralogy through images and rapid ground analysis methods. This provides valuable information on waste sites that cannot be gained through traditional visual inspection and field mapping methods. Rather than having to unnecessarily commit financial resources to extensive and slow field investigations, remote sensing data and analyses can be used to prioritize sites for detailed characterization and sample collection.

Remote sensing waste characterization techniques have proven applicable to both low-sulfidation mining districts (such as the Cripple Creek District) and high-sulfidation districts (such as the Goldfield District). Jarosite is the primary surface pathfinder mineral for identifying potential acid-generating materials (especially pyritic materials). Illite and smectites are of special interest for their metal absorption and releasing properties in districts where heavy metals could impact the downstream and surrounding environment. The mineral assemblages present at waste sites can be identified to varying degrees, depending on the type of remote sensing instrument used, with the most spectrally detailed methods allowing mineral analysis equal to, and in some ways better than, most laboratory X-ray and SEM methods.

Through knowledgeable application of remote sensing methods, waste inventorying and characterization will be more complete, faster, cost-effective, and unburdened by variations in field mapping methods and personnel which make existing surveys in the United States inconsistent (from state to state and agency to agency) and, therefore, of uncertain comparability and reliability. The methods described here can be applied as well to mined areas outside the United States, as evidenced by similar waste mineralogy found in many mining districts in many countries.

6.7
Applications of Imaging Spectroscopy Data: A Case Study at Summitville, Colorado

Trude V.V. King · Roger N. Clark · Gregg A. Swayze

6.7.1
Introduction

From 1985 through 1992, the Summitville open-pit mine produced gold from low-grade ore using cyanide heap-leach techniques, a method to extract gold whereby the ore pile is sprayed with water containing cyanide, which dissolves the minute gold grains. Environmental problems due to mining activity at Summitville include significant increases in acidic and metal-rich drainage from the site, leakage of cyanide-bearing solutions from the heap-leach pad into an underdrain system, and several surface leaks of cyanide-bearing solutions into the Wightman Fork of the Alamosa River. In general, drainage from the Summitville mine moves downslope into the Wightman Fork, a small tributary of the Alamosa River, which in turn flows east into the Terrace Reservoir before entering the agricultural lands of the San Luis Valley. The increase in the trace-metal burden of the Alamosa River watershed due to the mining activities at Summitville is of concern to farmers and fisherman, as well as Federal and State of Colorado agencies having responsibility for land stewardship.

The environment of the Summitville area is a result of 1) its geologic evolution, that culminated in the formation of precious-metal mineral deposits; and 2) previous metal mining activity. Mining accentuates, accelerates, and pertubates natural geochemical processes. The development of underground workings, open pits, mill tailings, and spoil heaps and the extractive processing of ore enhances the likelihood of releasing chemicals and elements to the surrounding areas and at increased rates relative to unmined areas. Both mined and unmined mineralized areas can produce acid drainage from the formation and movement of highly acidic water rich in heavy metals. This acidic water forms principally through the chemical reaction of oxygenated surface water and shallow subsurface water with rocks that contain sulfide minerals, producing sulphuric acid. Heavy metals can be leached by the acid solution that comes in contact with mineralized rocks, a process that may be enhanced by bacterial action. The resulting fluids may be highly toxic and, when mixed with groundwater, surface water, and soil, may have harmful effects on humans, animals, and plants. Thus, understanding the geologic and hydrologic history of this area is a critical piece of the environmental puzzle in the Summitville area.

The Summitville mine operators had ceased active mining and begun environmental remediation, including treatment of the heap-leach pile and installation of a water-treatment facility, when it declared bankruptcy in December 1992 and abandoned the mine site. The U.S. Environmental Protection Agency (EPA) immediately took over the Summitville site under EPA Superfund Emergency Response authority.

Summitville has focused public attention on the environmental effects of modern mineral-resource development. Soon after the mine was abandoned, Federal, State, and local agencies, along with Alamosa River water users and private companies, began extensive studies at the mine site and surrounding areas. These studies included analysis of water, soil, livestock and vegetation. The role of the U.S. Geological Survey (USGS) was to provide geologic, hydrologic and agricultural information about the mine and surrounding area and to describe and evaluate the environmental condition of the Summitville mine and the downstream effects of the mine on the San Luis Valley (King 1995).

6.7.2
Imaging Spectrometer Data

To address geological, hydrological, and agricultural problems, approximately 2 100 km^2 of imaging spectrometer data for the Summitville mining district, adjacent areas in the San Juan Mountains, and agricultural areas in the neighboring San Luis Valley were collected on September 3, 1992 (Fig. 6.74).

Imaging spectroscopy is a departure from traditional remote sensing concepts in that the data represent continuous, narrow-band spectral coverage over a selected portion of the electro-magnetic spectrum. Spectroscopic processing delineates absorption features in reflectance due to individual chemical bonds in surface materials and in the atmosphere and, when used with image analysis, maps their occurrence and distribution.

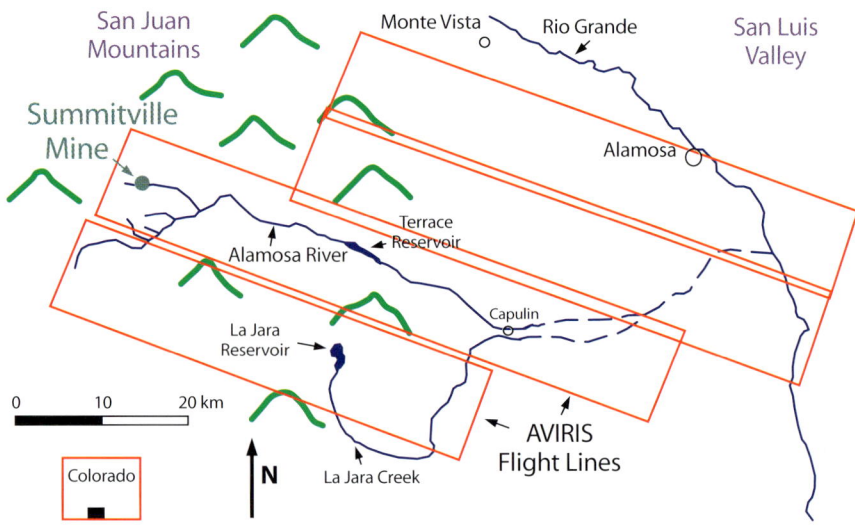

Fig. 6.74. Schematic map showing the location of the *AVIRIS* flight lines in the San Juan Mountains and San Luis Valley of Colorado

The narrow spectral channels of an imaging spectrometer form a nearly continuous sampling of the reflectance spectrum of the Earth's surface, in contrast with the 4 to 7 discontinuous broad channels of the other imaging instruments, such as *Landsat Thematic Mapper* (TM) and *Multispectral Scanner* (MSS). Imaging spectroscopy duplicates the capabilities of Landsat by distinguishing brightness and slope differences in the reflectance spectrum of the surface. However, imaging spectroscopy can also resolve absorption bands in the spectrum which can be used to identify spectral components. Analysis of imaging spectroscopy data allows minerals, vegetation types, manmade materials, water, snow, and many other materials to be mapped if they have unique and identifiable absorption features in the 0.4–2.45 µm spectral region (e.g. see Clark et al. 1992; Clark et al. 1993).

The imaging spectroscopy system used for this study was the *NASA* "Airborne Visible Infra-Red Imaging Spectrometer" (*AVIRIS*) instrument. *AVIRIS* acquires data in the spectral range from 0.4 to 2.45 µm in 224 spectral channels. The instrument is flown in an ER-2 aircraft (a modified U-2) at 19 800 meters (65 000 feet). The pixel size is 20 meters with 17 meters spacing between pixel centers; the swath width is about 10.5 kilometers (614 pixels). The swath length can be as much as 1 000 kilometers, if necessary, with the current tape recorder capacity on the aircraft.

A combined method of radiative transfer modeling and ground calibration were used to correct the *AVIRIS* radiance data to surface reflectance (Clark et al. 1995). This method, called Radiative Transfer Ground Calibration, or RTGC, corrects for variable water vapor and other components in the atmosphere, produces surface reflectance spectra free of unwanted artifacts, with spectral channel to channel noise approaching the true signal-to-noise ratio of the data. In the RTGC method, the initial step is to compute a radiative transfer model of atmospheric scattering and absorption for each pixel and remove these effects from the data. Secondly, a ground target is measured to characterize the spectral properties of a known area. The ground target characterization is used to correct any minor absorption residuals remaining after the radiative transfer model calculation and correction has been applied to the *AVIRIS* data. For this study, the primary calibration site was a plowed, vegetation-free field near the center of the *AVIRIS* data set. Approximately 12 individual samples from the calibration site were obtained on the day of the overflight and spectrally characterized on a modified Beckman-5 270 laboratory spectrometer (Clark et al. 1990b). These spectra then were averaged together to represent the spectral signature of the overall calibration site. The site soil samples were spectrally bland and, thus, suitable to use as calibration standards.

6.7.3
Data Analysis

Subsequent to the RTGC process, the data are analyzed to determine the materials present in the imaged scene. Clark et al. (1990a, 1991, 1992, 1995) developed a new analysis algorithm, Tetracorder, that uses a digital spectral library of well-characterized materials and a fast, modified-least-squares method of determining if spectral features for a given material are present in a pixel of the imaged data.

The current version of Tetracorder uses stepwise analyses that build on previous analytical results and directs subsequent analyses. In analyzing imaging spectroscopy data sets, many tens of materials can be found in a single scene and, perhaps, hundreds can be found if appropriate standards existed. Materials can be minerals, snow, vegetation (and different vegetation species), pollution, man-made objects, etc. Because so many different spectral signatures exist in a single scene, a spectral feature identification algorithm (Clark et al. 1990a) was developed as a first step in analysis. Using this feature matching algorithm, the analysis was expanded by Clark et al. (1991) to simultaneously analyze multiple spectral features in each unknown spectrum. Tetracorder applies multiple tests to match the spectrum under evaluation with standard spectra by comparing continuum-removed spectral features from the imaging spectrometer data set to corresponding continuum-removed spectral features from a reference spectral library (Clark et al. 1990b, Clark et al. 1991). The continuum removal process isolates an absorption band from "background absorptions" so that spectral features, including depth, shape, and wavelength position can be accurately analyzed and compared. To facilitate comparison, a straight-line continuum is removed from the library reference spectrum using data points (spectral channels) on each side of the absorption feature that is to be mapped. The same methods are used to remove the continuum from the observed spectrum in the flight image data. Several channels on each side of the absorption feature can be selected as continuum end-points to reduce noise in the continuum. The continuum is removed from the spectrum by division because of the non-linear effects of scattering and Beers Law. Adjacent spectral regions are analyzed simultaneously, resulting in multiple identifications when multiple spectral components are present. Depending on the results of this analysis phase, Tetracorder may choose additional analyses.

Other Tetracorder capabilities include fit, band depth, and continuum thresholding. For example, fit thresholding evaluates the value of the correlation coefficient from the least squares analysis. If the fit is too low to reasonably identify any of a given set of materials, the pixel is rejected from further analysis. Band depth thresholding works in a similar fashion. Continuum thresholding prevents material identification in pixels with very low continuum levels where noise could resemble absorption features. By knowing the continuum value, it is possible to determine if the spectrum represents an area in a shaded region, over water, or in a region obscured by a cloud.

The analytical methods take an all-inclusive approach to the observed electromagnetic spectrum, treating each spectrum as a single data object, rather than as a collection of 'channels' or discrete wavebands, thus allowing a more robust solution. Traditional analytical methods require a set of spectral end members (e.g., calcite, dolomite) to sum to one, which requires some of the end members to be identified in all pixels. Tetracorder analysis has no such requirement, and only finds materials if diagnostic absorption features are present.

6.7.4
Verification of Imaging Spectrometer Data and Results

Imaging spectrometer data allow several methods of verification, surpassing the capabilities of other remote sensing data. Common remote sensing data, such as *MSS* and *TM*, have a few spectral bands (data points) and each measured value can be influenced by a multitude of factors. Many factors, including, but not limited to, atmospheric absorptions, calibration errors, and surface material absorptions, affect the measured radiance, but the degree to which each factor influences the measurements cannot be determined. In contrast, a continuous spectrum over a wide wavelength interval (e.g. the ultraviolet to near-infrared wavelength region), such as from *AVIRIS* or a field-spectrometer, produces sufficient information to resolve the individual spectral contributions related to: (1) calibration errors, (2) atmospheric absorptions, (3) model artifacts, and (4) material surface absorptions. The spectral effects of each contribution then can be evaluated and corrected.

Data accuracy is a measurement of how well the imaging spectrometer data represent the true reflectance characteristics of the surface. Verification of the accuracy of imaging spectroscopy data, such as *AVIRIS*, can be accomplished by several means, including self-verification. In the self-verification method, spectra are extracted from the corrected image and examined. The wavelength calibration of the data set can be verified on the basis of known absorption features that do not shift in wavelength position (for example, atmospheric gases and selected materials). The correction of the imaging spectroscopy data to surface reflectance also can be confirmed using self-verification. Errors in the correction model or method used to derive the surface reflectance values can produce residual positive or negative spectral features that are recognized by an experienced spectroscopist as anomalous compared to the known spectral signatures of surface materials and atmospheric gases. Thus, the identified errors can be corrected to derive a viable surface reflectance.

Verification of imaging spectrometer material distribution maps, including minerals, vegetation, and man-made materials, can be accomplished by (1) traditional field verification methods (see below), and (2) by direct examination of the imaging spectrometer data. By extracting a spectrum from the data set, diagnostic absorption features can be identified and material identification confirmed based on their wavelength position and shape. This direct examination of the image can be done without field checking. This latter method only works for those materials having diagnostic absorptions that are separated from the spectral features of material mixtures. In many instances, the individual components of material mixtures can be identified even if overlapping absorption features exist. By visually examining a spectra or by using modern spectral identification algorithms, individual minerals and mixtures, such as kaolinite and hematite, montmorillonite and goethite, jarosite and muscovite, calcite or dolomite, can be positively identified. However, some material mixtures can produce absorption features that can be confused with other materials (e.g., a kaolinite and smectite mixture is spectrally similar to a spectrum of halloysite). Consequently, field checking is often required to properly determine the composition.

Traditional field checking methods require a sample from a specific geographic area represented by a pixel in the image. The sample then is analyzed by a method (e.g. X-ray diffraction or a micro-beam technique), that provides (1) mineralogy, (2) mineral composition; and (3) mineral abundance. These data then are compared to the spectral information. Given that the imaging spectroscopy data represent only the surface of the exposed material, care must be taken during the laboratory phase to ensure that the materials analyzed in the laboratory are the same as those mapped by the imaging spectrometer. Different analytical techniques have different levels of detection for specific materials. For example, spectroscopy is more sensitive for the identification of clays, iron oxides, and amorphous materials, sometimes by factors of ten compared to X-ray diffraction methods. In contrast, X-ray diffraction analysis commonly can discriminate between kaolinite and smectite mixtures, or kaolinite and muscovite mixtures, and halloysite, which are difficult to discriminate using spectroscopy data. Failure to evaluate the detection limitations of each method can result in a disagreement between different analytical techniques. Field-checking verifies both the accuracy of the mapping algorithm and the spectral standards and is the most rigorous type of validation. Unfortunately, field-checking can be time-consuming, costly and, in some instances, may not be logistically feasible.

6.7.5
Mapping Minerals

Reflectance spectra can provide information on the chemistry, mineralogy and crystal structure of materials. Spectroscopy can identify both crystalline and amorphous materials. Subtle changes in composition or structure will result in changes in shape and/or wavelength position of characteristic absorption features. Laboratory investigations have quantified the causes of changes in position and shape of absorption features for particulate minerals (Hunt 1977 and references therein, Adams 1975, Gaffey et al. 1993), carbonates (Gaffey 1986, 1987), silicates (Cloutis et al.

1986, Cloutis and Gaffey 1987; King and Ridley 1987), and phyllosilicates (King and Clark 1989, Clark et al. 1990a).

Advancements in remotely-sensed data acquisition platforms and computer analysis techniques now allow a direct comparison between laboratory and remotely sensed data. Thus, well-characterized laboratory spectral databases can serve as "standards" to map the spatial distribution and composition of remotely-sensed materials.

Reflectance spectroscopy identifies and maps specific chemical bonds in materials which are present in the upper 1–2 mm (in most instances) of an exposed surface. Mineralogically, this material commonly represents the layer which has been exposed to physical and chemical weathering. These minerals may all be secondary weathering products, depending on the composition of the parent mineral/rock and the physio-chemical conditions of weathering. The ability to define and map secondary minerals and amorphous, or poorly crystalline, materials is important because they are common and are sources of easily-available trace metals and anions. Thus, by combining the spatial distribution and mineral chemistry data gained from the imaging spectroscopy data, with supporting field studies, it is possible to identify sources of pollution, monitor mineral transport and fate, predict metal geoavailabitlity, and assess mineral associations.

Absorption bands in the visible and near-IR portion of the spectrum (~0.4–1.0 µm) are caused by electronic processes, including those due to crystal field effects, charge transfer, color-centers, and conduction bands. The absorption features resulting in this portion of the spectrum often involve elements of the first transition series, which have an outer unfilled electron orbital d-shell. The energy levels are determined by the valence state, coordination number, and site symmetry of the element. Differences in these parameters between materials are manifested as individual diagnostic absorption bands in the visible and very near infrared wavelength regions. Absorptions in this wavelength region commonly are associated with the presence of iron (Fe) and other transition elements (Mn, Cr, Ti, etc.) in the mineral structure. The intrinsic strength of these absorptions is quite strong. Therefore, reflectance spectroscopy is well suited to study the varied Fe-bearing oxides, sulfates, and hydroxides produced by typical weathering associated with unmined and mined mineralized areas.

Near-infrared radiation (1–2.5 µm, in this study) absorbed by minerals and other materials is most commonly converted into molecular vibrational energy. The frequency or wavelength of the absorption depends on the relative masses and geometry of the atoms and the force constants of the bonds.

Absorption features in the 2.2 to 2.3 µm region are commonly used to make mineral identifications. Many of the absorption features in this wavelength region result from a combination of the OH-stretching fundamental with either the Al-O-H bending mode (usually absorbing at approximately 2.2 µm), the Mg-O-H bending mode (absorption usually near 2.3 µm), and to Fe-O-H bending mode absorbing near 2.25 µm. At high spectral resolution, these bands are recognized to be complex absorption features. Based on previous work (King and Clark 1989, Clark et al. 1990b, Swayze and Clark 1990, Clark et al. 1993, and others), the strength, position and shape of these features has been found to be a function of the mineral chemistry. Similarly, an overtone of the fundamental asymmetrical stretching mode of C-O in the carbonate ion produces an absorption feature in the 2.3 to 2.34 µm wavelength region.

For the present study, minerals were mapped based on the presence of absorption features in the ~0.45 to 2.45 µm wavelength region (i.e., the visible and near-infrared portions of the electro-magnetic spectrum). A laboratory standard spectral database totaling 130 minerals was used as a basis for comparison with the remotely sensed data. The database included pure minerals, mineral mixtures, and materials collected from the Summitville study area. From this database, a search was conducted for 50 minerals with absorption features at wavelengths near or less than 1.0 µm and 14 of these 50 potential minerals were mapped in the Summitville/San Luis Valley region. The 14 mapped materials are predominately crystalline and amorphous iron-oxides, thus their spectral identification was based primarily on the wavelength position and shape of the continuum removed 0.9 µm absorption feature. Continuum removal, as previously mentioned, allows subtle differences to be distinguished between similar, but distinct, absorption features. The minerals detected include: two types of goethite, two types of jarosite, two different Fe-bearing minerals, nanohematite, coarse- and fine-grained hematite, amorphous Fe-oxide, ferrihydrite, wet amorphous Fe-oxide, Cu-bearing precipitate, and sediment-bearing water.

To evaluate the presence of minerals that have absorption features in the 2.2–2.3 µm wavelength region, 80 laboratory mineral standards were used. These standards included phyllosilicates, sulphates, carbonates, and cyanide compounds. Eleven (11) different minerals of significant areal extent were detected in the *AVIRIS* data. Subtle spectral differences allow discrimination between K and Na alunites and between poorly-crystalline and highly-crystalline kaolinites. However, because of spectral similarities and limitations of the mapping algorithm, some of the material mapped as Na-montmorillonite may be muscovite or sericite.

Spectral data contained in the *AVIRIS* pixels are very similar to the spectral standards measured in the laboratory. Figure 6.75 shows the spectrum of a mixture of alunite, kaolinite and goethite detected in the *AVIRIS* data from near Alum Creek compared to laboratory standard spectra of an alunite and kaolinite mixture and goethite.

Breaks in the spectra of the *AVIRIS* data occur at the wavelengths where absorptions from atmospheric gases occur. The absorption features in the spectrum of the mineral mixture from the *AVIRIS* data can be correlated with the diagnostic absorption features in the laboratory standard. The absorption features near 0.7 and 1.0 µm in the pixel spectrum result from the presence of iron in goethite. The absorption features between 1.4 and 1.7 µm and the one near 2.3 µm are due to the presence of alunite. Those near 2.2 µm result from the presence of kaolinite in the pixel spectrum.

Spectral processing and analysis of *AVIRIS* data using absorption features in the 2 µm wavelength region identified areas of hydrothermal alteration minerals. Data covering the Summitville mine showed mineralogical differences between the open-pit and the heap-leach-pile at the mine site. Hydroxyl-bearing materials, including clays, show discrete distribution patterns at both the mine site and within the Iron Creek, Alum Creek, and Bitter Creek basins (Fig. 6.76). The data also defined discrete mineralogical boundaries in one other basin (Burnt Creek) which is the principle drainage for mineralized regions that are not included in the over-flight area.

An interesting observation is that the Summitville mine, at the time of the over-flight, apparently did not contribute OH-bearing minerals via the Wightman Fork

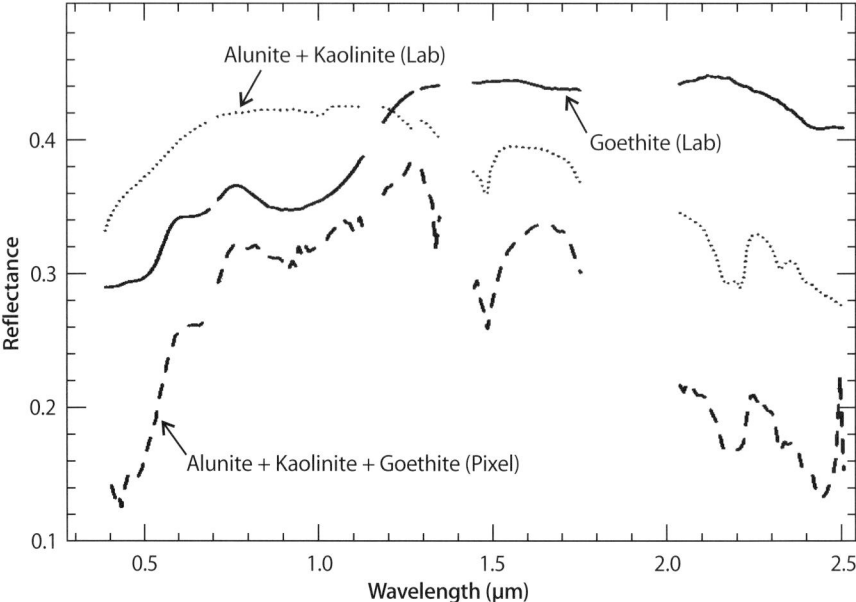

Fig. 6.75. Spectra of muscovite and goethite measured in the laboratory and muscovite and goethite mixtures detected by *AVIRIS* near Alum Creek. Note the match between the *AVIRIS* pixel and goethite in the 1 μm region and the laboratory muscovite and the *AVIRIS* pixel in the 2 μm wavelength region. The gaps in the *AVIRIS* spectrum are wavelengths where atmospheric gases absorb

to the Alamosa River. In contrast, the mineralized areas in Iron Creek, Alum Creek and Bitter Creek basins do contribute OH-bearing minerals to the Alamosa River. This observation is based on the spectral characteristics of the exposed fluvial sediments along Alum Creek and Bitter Creek and lack of exposed OH-bearing fluvial sediment along the Wightman Fork. The unmined mineralized areas are believed to contribute OH-bearing materials to the Alamosa River due to the porous character of the well-exposed hydrothermally altered bedrock, which allows altered materials to be eroded easily and carried downstream. The paucity of phyllosilicates being transported from the mine site to the Wightman Fork may be attributed to on-site remediation efforts. Conversely, the material may be carried as a very fine-grained aqueous suspension, which does not settle onto the creek banks, or as aqueous components that may precipitate under higher pH conditions downstream.

From the *AVIRIS* data, it appears that only a small amount of material having vibrational absorptions is being transported from all upstream locations into the Terrace Reservoir via the Alamosa River. At the time of data acquisition, no material (or amounts below the spectral detection limit) having absorption features near 2.2–2.3 μm were being discharged from the Terrace Reservoir. Analogously, there is no spectral indication of alteration materials having 2.2–2.3 μm absorption features being transported via La Jara Creek to the San Luis Valley (Fig. 6.74).

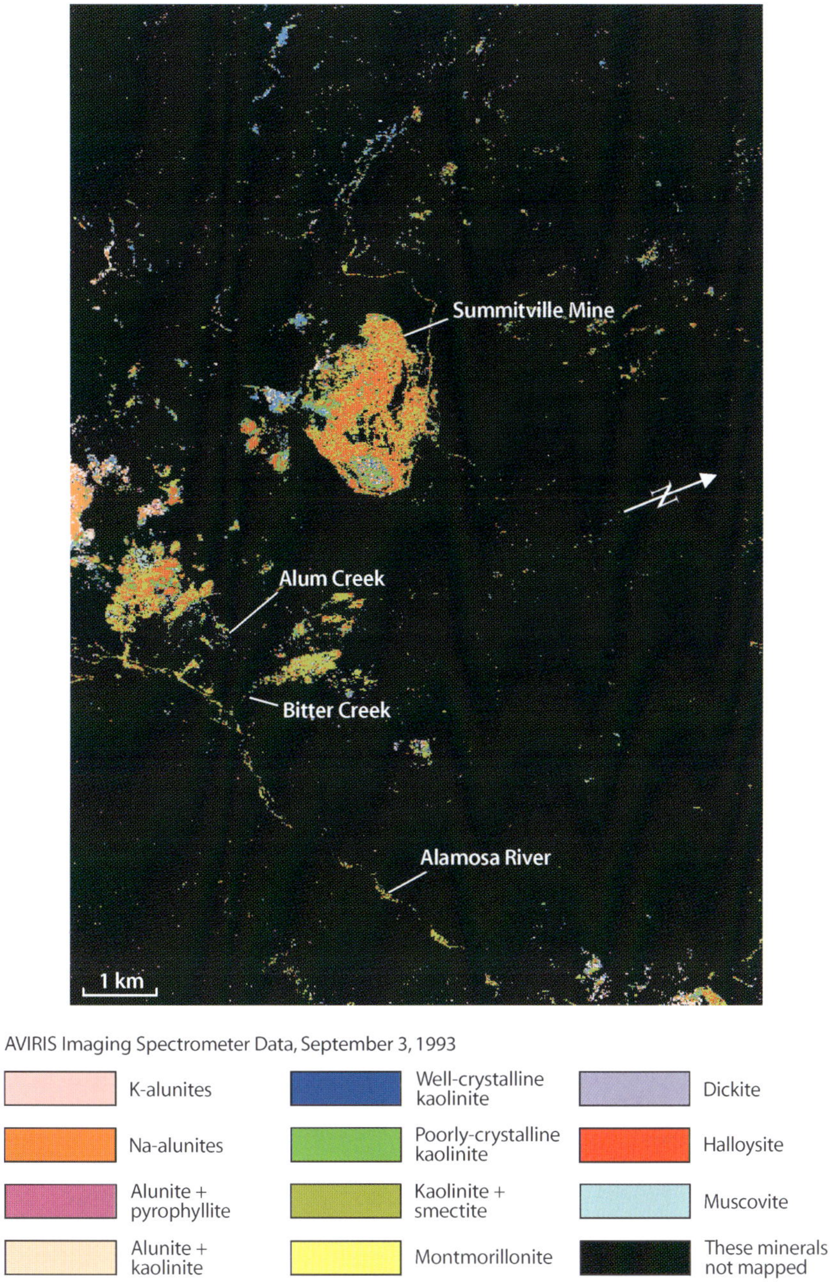

Chapter 6 · Case Studies

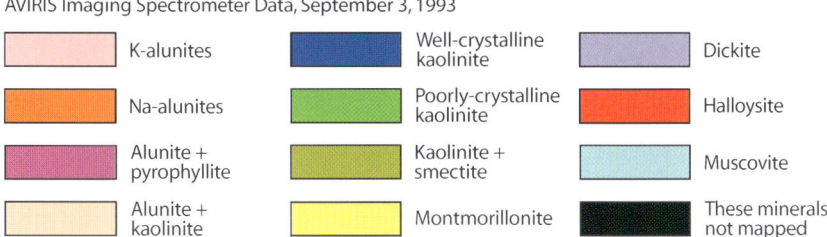

Fig. 6.76. *AVIRIS* image of the Summitville mine and surrounding area showing the distribution of minerals having absorption features in the 2.0–2.4 μm wavelength region. The image covers an area from approximately 37°22'30"N to 37°27'30"N latitude, 106°30'W to 106°37'30"W longitude

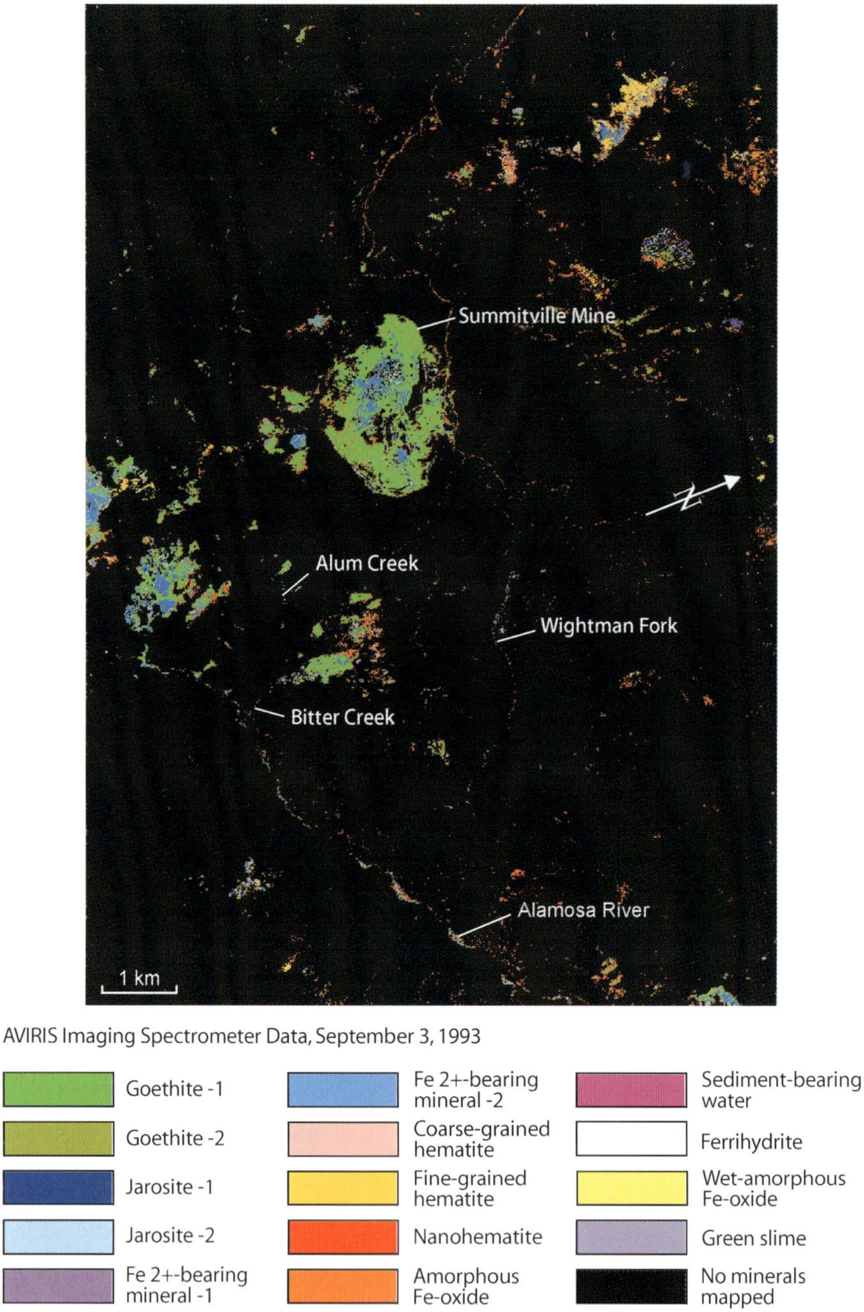

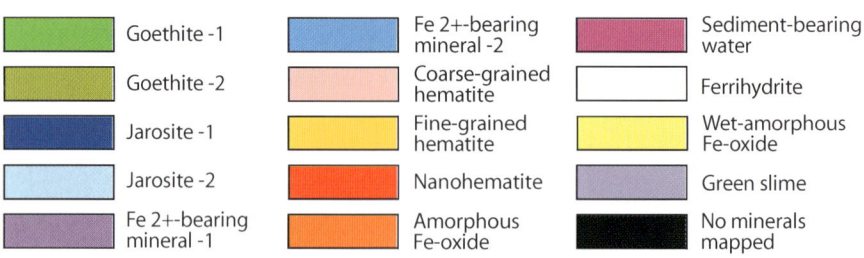

Fig. 6.77. *AVIRIS* image of the Summitville mine and surrounding area showing the distribution of minerals having Fe-absorption features. The image covers an area from approximately 37°22'30"N to 37°27'30"N latitude, 106°30'W to 106°37'30"W longitude

Spectrally detectable quantities of OH-bearing materials are present in some plowed and/or harvested fields in the San Luis Valley. On-site field-checking of the plowed fields revealed small to fist-size clumps of alunite, suggesting that at some earlier time the alunite and associated minerals were transported from areas of exposed altered material, presumably near the Alum Creek, Bitter Creek, or Iron Creek basins, to the valley floor.

AVIRIS images show that the Summitville mine and Iron Creek, Alum Creek, and Bitter Creek basins are sources of iron (Fe)-bearing sediments to the Alamosa River (Fig. 6.77). Field observations show that these sediments give a reddish-brown color to stream banks, a characteristic typically associated with acid drainage, and are potential carriers of heavy metals to locations downstream. The *AVIRIS* data show that the Fe-bearing materials enter the Terrace Reservoir via the Alamosa River. Consequently, in assessing the environmental impact of mining near Summitville, it is important to recognize that both the Summitville mine site and the local drainage basins associated with unmined mineralized areas are contributors of Fe-bearing sediments and aqueous components.

Based on the *AVIRIS* data analysis, there are no other sources of alteration minerals from mineralized areas of substantial size that significantly influence the distribution of Fe-bearing sediments. However, two large alluvial fans associated with the Alamosa River and La Jara Creek distribute Fe-bearing sediments to the Valley floor (Fig. 6.74). Both of these alluvial fans show distinct lobes, some of which have been incorporated into cultivated fields. The wide-spread aerial distribution patterns of the Fe-bearing sediments incorporated in the fans indicates that they are the natural weathering products of the volcanic rocks in the San Juan Mountains and, for the most part, are not related to mining activities at the Summitville mine.

Reservoirs and lakes (including La Jara and Terrace Reservoirs) in the data set have been mapped as specific minerals, which is inaccurate. However, based on spectra that have been extracted from the remotely sensed data set, these water bodies are different than our standard water spectra. The La Jara Reservoir and some lakes (Big Lake) in the data set map similarly, but the Terrace Reservoir appears anomalous in mineral maps. The *AVIRIS* data may be detecting suspended sediments, algae, or bottom sediments, but the exact character of the materials detected in the bodies of water and why they map as minerals are subjects of ongoing studies.

6.7.6
Mapping Vegetation

Obtaining quantitative information about vegetation has proven difficult. To first order, all vegetation is chemically similar, and most healthy plants are green and have similar absorption features. However, the human eye sees plants as shades of green, thus allowing subtle, but significant, spectral differences between plant species to be distinguished. Quantifying these differences through spectroscopy provides the ability to map plant species, determine the water abundance in a species, and determine the relative health of a species or community.

The primary spectral features used for vegetation identification result from the presence of chlorophyll, the organic material that gives plants their green color.

The human eye sees different plants as shades of green, because of the "green peak" of reflected light. The "green peak" results from the absorption of most wavelengths of light, except those that appear as green to the human eye, in the visible portion of the electro-magnetic spectrum. The broad absorption features that are responsible for the overall similarity in the green color of vegetation spectra result from the presence of chlorophyll and other pigments. However, absorptions due to bending and stretching of the O-H bond in water, as well as the presence of carbon and nitrogen in the plant structure, result in absorption features at other wavelengths as well (Danks et al. 1984; Murray and Williams 1987, Curran 1989, Curran et al. 1992).

Although the spectra of plants are sufficiently different to allow species identification, the spectra of an individual species can vary. Spectral variations probably result from the amount of chlorophyll and water in the plant, a complex relation between the stages of a growing cycle and the health of a plant. The health of a plant can be affected by many factors, including the amount of water available (too much or too little) and metal toxicity. These factors influence the shape and depth of the characteristic absorption features.

The Tetracorder algorithm is very sensitive to the shape of spectral features and has the potential to distinguish more subtle differences in the visible spectrum of plants than the human eye. The continuum-removal portion of the algorithm is an important step in detecting and mapping vegetation, particularly when a pixel contains spectral information from green plants, dry vegetation, and soil. By isolating the absorption features with continuum removal, the position and shape of the continuum removed-spectral feature remains constant, although its depth changes as a function of the areal extent of the vegetation in the pixel.

The extreme changes in elevation from ~3 960 m near the Summitville mine to ~2 300 m in the San Luis Valley, is reflected by diverse vegetation communities, ranging from alpine to irrigated agricultural environments. Much of the mapped area is above tree-line at the higher elevations, and grades downward into mixed Lodgepole and Ponderosa Pine, Douglas Fir, Aspen and deciduous growth at intermediate elevations.

At the lower elevations, the study area includes farmland producing potatoes, alfalfa, barley, oat hay, canola, and fields containing chico and other unidentified weeds. Ideally, data analysis could use a digital spectral library of reference spectra for all plant species likely to be encountered in the study area. However, such a library does not exist for vegetation as it does for minerals (e.g. Clark et al. 1993). Producing a spectral library for vegetation will be much more complex than for minerals, as the vegetation is not static; stress factors and the number of spectra, as a function of growing season, that would be required to adequately represent a plant species is not known.

Reference spectra were obtained for sites of known species directly from the *AVIRIS* data. *AVIRIS*, having been well calibrated to surface reflectance, acts as an excellent field spectrometer, providing data for large areas and averaging over many plants to reduce spectral variations within one species.

The reference spectra obtained from the *AVIRIS* data set are shown in Fig. 6.78. The alfalfa, canola, oat hay, and Nugget potato spectra (Fig. 6.78a) show the plants to be green and healthy. The barley had lost all its chlorophyll signature because of natural senescence (Fig. 6.78b). The Norkotah potatoes were not being irrigated

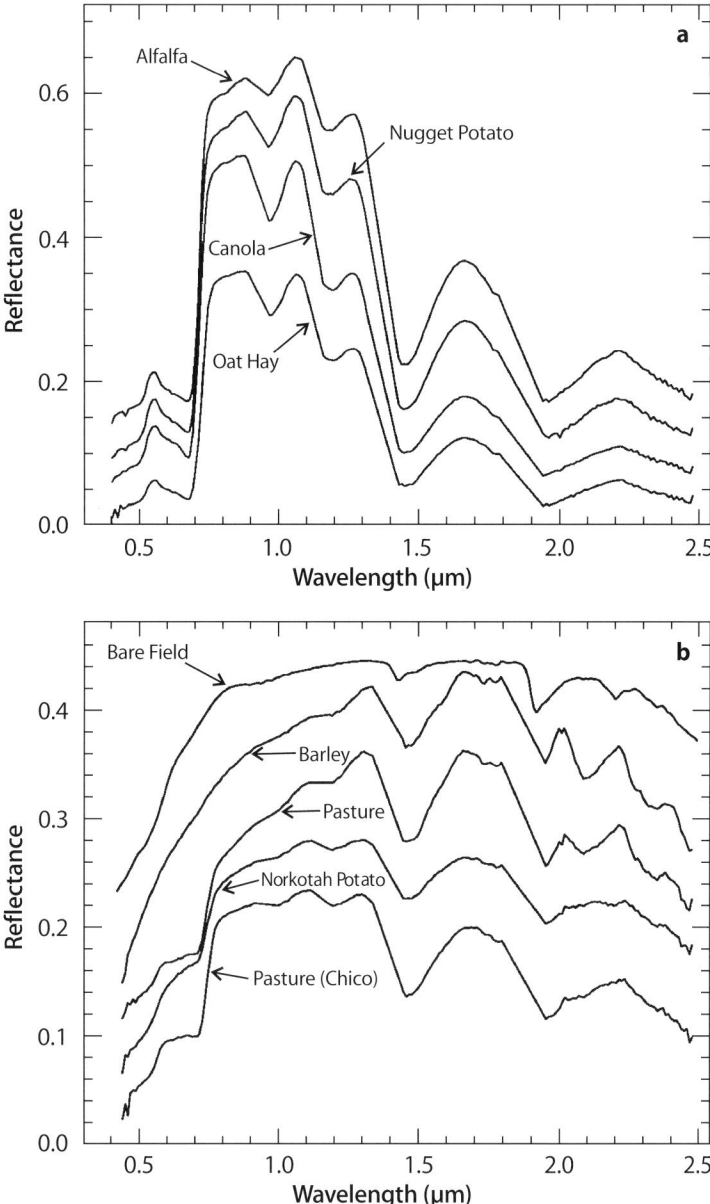

Fig. 6.78. Reference spectra used in the mapping of vegetation species. The field calibration spectrum is from a sample measured on a laboratory spectrometer; all others are averages of several spectra extracted from the *AVIRIS* data. Note that the noise is extremely low, comparable to the lab spectrum of the field calibration site. In **a**, each curve has been offset from the one below it by 0.05. In **b**, each spectrum has been offset by 0.04 from the one below it, except the top spectrum is offset 0.06. The offsets are cumulative, so the field calibration spectrum is offset a total of 0.18 for clarity

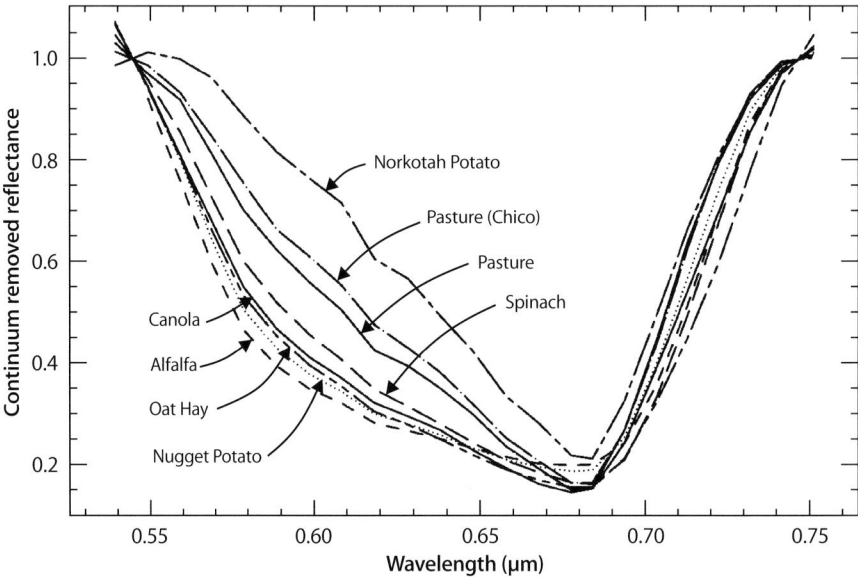

Fig. 6.79. The continuum-removed chlorophyll absorption spectra from Fig. 6.78 are compared. Note the subtle changes in the shapes of the absorption between species

as they were about to be harvested, and consequently they showed a weak chlorophyll and cellulose absorptions as well as clay absorptions from exposed soil. These potatoes were also being sprayed with a defoliant in preparation for harvest and should show decreased chlorophyll absorption, along with a shift of the red edge of the absorption to shorter wavelengths due to drying. The chico and pasture spectra show combinations of chlorophyll and cellulose (dry vegetation) absorptions, which can be attributed to seasonal and species variability.

Differences in the shape of the absorption features in the continuum-removed (chlorophyll-containing) crop spectra enabled differentiation using the *AVIRIS* data (Fig. 6.79). The continuum-removed spectrum of each pixel in the image was examined and compared to the standard crop spectra to produce a color-coded distribution map (Fig. 6.80). Farmers in the San Luis Valley rely heavily on irrigation waters applied by center-pivot irrigation systems, thus most farmed lands appear as circular areas in the *AVIRIS* data.

Field verification data were supplied by Maya ter Kuile of Argo Engineering (personal communication 1993), and as part of this investigation. Of 43 verification fields (not including chico/pasture areas), 7 included the sites of reference spectra and were, of course, identified correctly. Of the remaining 36 fields, 33 were identified correctly and another 3 were identified as mixed by the Tetracorder analysis (but were indicated as one crop type in the field data); no fields were misidentified. The mixed fields were identified as consisting of two species, one of which was correct in each case. If a score of half is given to the three fields identified as mixed, the 96% score (34.5 of 36) implies the method is accurate.

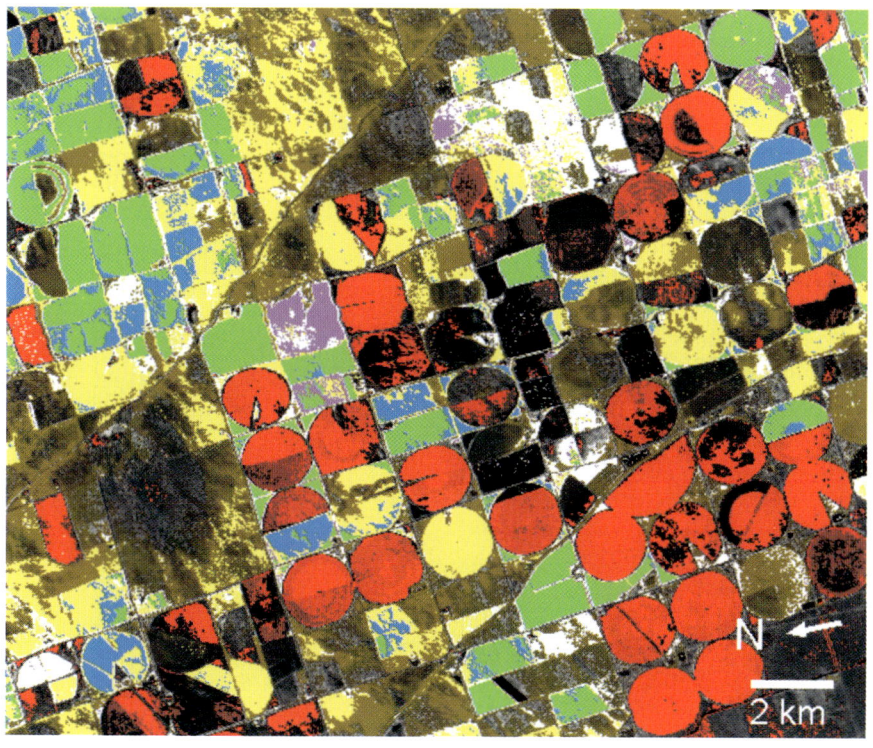

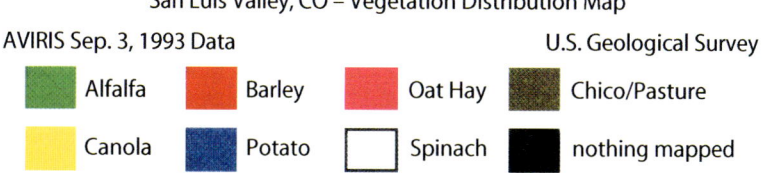

Fig. 6.80. The Tetracorder analysis of vegetation species (see text)

The accuracy is even more impressive considering several fields were mapped correctly even though they had already been harvested suggesting that low concentrations of plant residue are sufficient to make correct species identification. The harvested areas are identified in Fig. 6.80 by circular plots where the colored pixels are sparse and/or low in intensity. The spectra of the harvested areas are also identified in the senescence/stress map (see below).

Canola was mapped in many of the areas known to be chico/pasture. While it is possible that canola seeds have been blown into surrounding fields and are thriving, it is more likely that the canola spectral signature is similar to other plants in those uncultivated fields for which standard spectra were lacking.

The distribution of vegetation species/communities based on the shape of the chlorophyll absorption features near the Summitville mine are shown in Fig. 6.81.

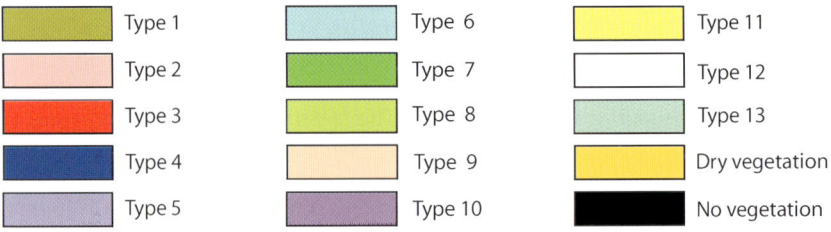

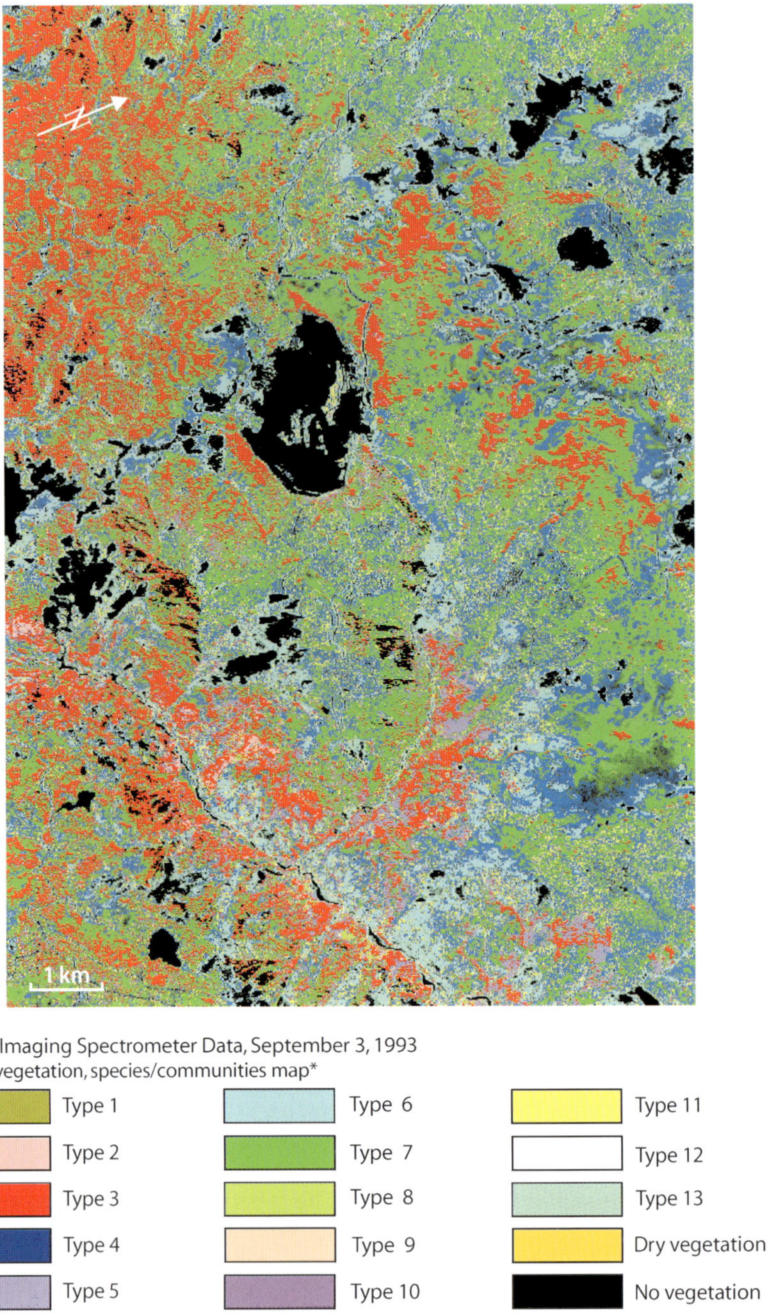

Fig. 6.81. The Tetracorder analysis of vegetation species/communities near the Summitville mine based on the shape of the chlorphyll absorption (* The same colors at different elevations may by different species/communities)

Fig. 6.82. The Tetracorder analysis of vegetation near the Summitville mine showing relative dryness of the species based on the 0.95 and 1.2 μm absorption features

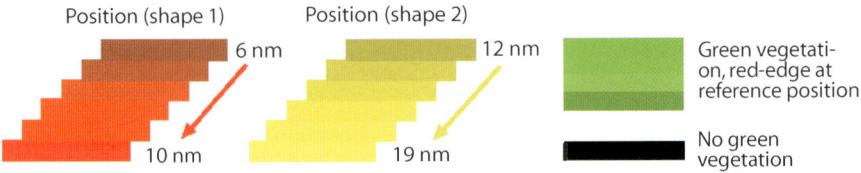

Fig. 6.83. The Tetracorder analysis of vegetation from near the Summitville mine showing differences in the senescence/stress levels

Fifteen generic vegetation types are identified within this *AVIRIS* scene. However, because of a paucity of standard reference vegetation spectra for this region and the rapid onset of winter after the data were collected, supporting field-work was not an option. Consequently specific species names can not be given for this portion of the study. Thus, Fig. 6.81 is indicative of vegetation communities rather than individual species. It is interesting to note the differences in communities on north and south facing slopes. These differences may be because of differences in community distribution or indicative of differences in fall-colors (natural senescence) between north and south facing slopes or both.

Figure 6.82 shows a green-vegetation/water-abundance map for the area near Summitville, based on the depth of 0.95 and 1.2 μm water absorption features and the chlorophyll absorption feature. These features are indicative of the dryness of vegetation in the *AVIRIS* data. Low abundances of chlorophyll or water, as indicated by either the 0.95 or 1.2 μm water absorption features or chlorophyll absorption, can produce similar spectral affects. If water abundances are low, the mixing of the three end member colors, representing the three absorption features, will produce red to reddish-brown colors as seen on Fig. 6.82. Because the data were acquired in September, the natural senescence of the vegetation at the higher elevations had begun and preferential "drying" was taking place on the south-facing slopes. Comparably, the agricultural crops in the San Luis Valley show different stages of maturation and, consequently, dryness.

6.7.7
Senescence/Stress Mapping

The long-wavelength side of the chlorophyll absorption (~0.68 to ~0.73 μm) forms one of the most extreme slopes found in spectra of naturally occurring common materials, plants or minerals. The absorption is usually very intense, ranging from a reflectance low of less than 5% (near 0.68 μm) to a near infrared reflectance maximum of ~50% or more (at ~0.73 μm). The properties of the reflectance spectra of plants often indicate that this absorption band is "saturated". If the absorption feature is saturated, the wavelength position of the absorption band minimum will not differ significantly, but the wings, and consequently the width, of the absorption will change. When the chlorophyll absorption in the plant decreases, the overall width of the absorption band decreases. The short wavelength side of the chlorophyll absorption can not be observed in reflectance spectra because of other absorptions in the ultraviolet (UV) wavelength region. In spite of these observational limitations, decreases in the strength of the chlorophyll absorption (less chlorophyll in the plant) causes a shift of the long wavelength side of the absorption feature toward shorter wavelengths. This has popularly become known as the "red-edge shift" or the "blue shift of the red edge" and has been used by researchers as an indication of senescence or stress-induced chlorosis (e.g. Milton et al. 1983, Rock et al. 1985, Miller et al. 1987, Milton et al. 1989). However, questions remain about the limits of these measurements for canopies and leaves, as well as the exact causes of these shifts (Curran et al. 1991).

To determine if a red-edge-shift occurred, where it occurred, and the relative amount of shift within the data set, we used field spectrometer spectra from the

San Luis data set to compute a ratio cube. The ratio of two spectra, each having steep "red edge" spectral slopes, which are shifted in wavelength relative to the other, will produce a spurious value even if there is only a small relative shift between them. If a "blue" shifted spectrum is divided by an unshifted spectrum, a peak will be observed in the ratio. For a spectrum of green vegetation (from Fig. 6.78), a 1 nm shift will produce a residual feature of approximately 6% (Clark et al. 1995). The *AVIRIS* data have a signal to noise ratio of several hundred in this spectral region, so red-edge shifts of less than 0.1 nm can be detected. Figure 6.83 shows differences in the senescence/stress of the vegetation from the Summitville mine site to the San Luis Valley based on differences in the position and shape of the chlorophyll absorption feature.

Many red-edge shifts are observed in the data set (Fig. 6.83). However, to be useful as an indicator of stress, the red-edge-shift map should be compared to the community/species map to reduce interpretative bias, as red-edge-shift can vary as a function of species. Comparison between these two maps suggests that no major red-edge shifts can be related to mining activities. However, the time of the data collection (September) was not optimum for studies focusing on the spectral detection of vegetation stress. Natural senescence and agricultural processes (defoliation) could conceal the spectral identification of plant stress related to the uptake of toxic materials. It should be noted that if an area of "stressed vegetation" was suggested in the *AVIRIS* data, additional field work and analysis would be necessary for confirmation.

6.7.8
Conclusions

The unique utility of imaging spectroscopy in mapping mineral and vegetation distribution on both local and regional scales at the Summitville mine and adjacent portions of the San Juan Mountains and San Luis Valley has been demonstrated. Imaging spectroscopy data provides mineralogical and chemical data unavailable from any other remote sensing method, and is an excellent tool for environmental assessments, mineral mapping and exploration, vegetation communities/species and health studies, and general land management applications. For geologic studies, imaging spectroscopy data can be used to detect and map sources, pathways of transport, and fate of materials in an area having unmined hydrothermal alteration and mining activities. In addition, applications of imaging spectroscopy data for several aspects of botanical investigations have been demonstrated.

In the Summitville mining region, the mine site does not contribute clay minerals to the Alamosa River, but does contribute Fe-bearing minerals, and the Fe-bearing materials are being transported into the Terrace Reservoir. Such minerals have the potential to carry heavy metals and their transport and fate need to be monitored as they can serve as a source of contamination. The imaging data suggest that if hydroxyl-bearing and sulphate-bearing materials are being transported in the Wightman Fork, they are in suspension and have not been deposited in bank-materials in this reach of the Summitville watershed, but may be deposited further downstream due to variations in water chemistry.

The *AVIRIS* data suggest that alteration materials, including alunites, montmorillonites, and kaolinite/smectites have previously been transported to the valley floor from the area near the Summitville mine and/or Alum Creek, Iron Creek, or Bitter Creek basins as suggested by the spectral signature of some plowed fields. Alluvial material has been widely dispersed on the Valley floor from both the Alamosa and La Jara drainages. Because of its widespread distribution and the lack of either unmined mineralized or mining sources in the La Jara basin, these materials are believed to be primarily related to the natural weathering of the volcanic rocks in the adjacent San Juan Mountains and are not environmentally significant as sources of heavy metals.

The utility of imaging spectroscopy as a source of botanical data for environmental applications, to monitor vegetation cover and its health, and define the distribution of vegetation communities and specific species has been demonstrated. The data presented here illustrate the advantage of using continuum-removed spectra for general detection and mapping of agricultural crops, as well as defining differences between closely related species. The ability to remotely detect and map differences in plant species could be used for more specific, and potentially more accurate, crop yield predictions.

The *AVIRIS* data for the Summitville area and San Luis Valley show that some vegetation spectra have a shift in the wavelength position of the chlorophyll absorption feature (a red-edge-shift). However, when compared to vegetation community maps, these changes in the indigenous vegetation and agricultural crops cannot be attributed to metal loading by either the mined or unmined mineralized areas in the region.

Epilogue

In the first part of this volume, we have provided background information on current remote sensing technology. The techniques and instruments in use span the entire range of levels of technical difficulty, from aerial photographs and their visual interpretation to airborne hyperspectral imaging spectrometers and the attendant intensive computer processing of their data. All levels of technology and types of output information have their place in site characterization work.

As seen in the case histories and examples given for different sites, aerial photography provides the most complete historical record for most sites when available. Paper maps of pre-1970s waste sites often do not exist or are incomplete. Organized aerial photography programs in most industrialized countries began in the 1930s or 1940s, with some coverage extending back to the 1910s (taken from early powered aircraft and balloons) or even earlier in isolated cases (such as photographs taken from observation balloons during the U.S. Civil War). Satellite photography did not start until the 1950s (after the start of the Space Age) and did not become publicly available until the 1970s for the most part. Most available photos are black and white, thereby placing constraints on their use for identifying material and vegetation differences. More modern color and color infrared photographs provide much more useful information on vegetation and some material differences in and around sites of interest.

The need to identify finer differences in vegetation, mineralogy, thermal emittance, and other ground characteristics, for multiple purposes, led to the development of satellite and airborne multispectral devices that could record ground reflectance information in parts of the electromagnetic spectrum previously inaccessible to most photographic systems. Information from the near-infrared portion of the spectrum allowed development of new techniques for vegetation analysis (vigor, quantity, stress, etc.) and for spectral analysis of mineral groups that allowed remote geologic mapping. These devices have progressed over the past three decades to the point where airborne hyperspectral devices (and future satellite devices) allow mapping of individual minerals and plant species (and stresses) to a degree that previously could only be performed with laboratory analyses.

Thermal remote sensing techniques are very useful for identifying surface and groundwater properties and locations. Likewise, these techniques can provide information on surface effects of subsurface conditions (such as thermal anomalies due to voids or thermally active parts of wastes) that otherwise may not be detectable by ground or remote sensing investigation methods. Other technology, such as radar interferometry or laser ranging devices, can provide information on topography and terrain changes much more efficiently and completely than ground surveys and photographic methods. These were not covered in detail in this volume because of their relatively recent addition to the panoply of remote sensing methods and their limited use as yet for environmental site characterization.

Field work remains an indispensable part of remote sensing site characterization work. Site visits must be performed at least on a spot basis to determine that what has been identified through remote sensing is indeed what is on the ground and to identify other factors that may not have been evident from just remote sensing data. Spot sampling (and often subsequent laboratory analysis) of materials at the site is a normal part of these investigations. Use of field spectrometers, when appropriate such as for mineralogical and vegetation studies, is an important aspect of ground truthing and for the collection of information on the site that can be used to better analyze the airborne or satellite remote sensing data, both for the site in question and for application of results to other similar sites.

Of course, investigation of the subsurface properties and conditions of waste sites is very limited to impossible through remote sensing techniques alone. Subsurface investigation via geophysical and hydrological techniques and drilling must be performed to provide any certainty of how surface properties identified through remote sensing relate to other site characteristics (such as types of buried wastes). The second volume (to be published in 2000) in this Methods in Environmental Geology series addresses geochemical techniques for site characterization.

The case histories presented in this volume do not, of course, cover all possible applications of remote sensing to site characterization, either in terms of types of remote sensing technology or types of waste sites. However, this volume should provide managers, scientists, and applications personnel with a solid background on most advantages and limitations of available remote sensing techniques.

We have provided the basis for understanding applied remote sensing and successful use of this technology for site characterization. Now it is up to you, the environmental manager and professional, to take this knowledge and apply it to your own sites and investigations so that you may more thoroughly evaluate waste sites of all kinds. Proper use of remote sensing in site studies will improve the timeliness and cost effectiveness of characterization work and will benefit clients, the general public, and our environment by providing better information for remediation planning and performance!

Acknowledgements

The authors and editors would like to thank the German Federal Ministry of Education, Science, Research, and Technology (BMBF), the project sponsor, the Federal Environmental Agency (UBA), the United States Geological Survey (USGS) and the project management of "Waste Disposal Underground" in the Federal Institute for Geosciences and Natural Resources (BGR) for their support.

This book has much benefitted from helpful comments and reviews by Dan Knepper (Denver USGS).

Bernhard Hoerig, Friedrich Kuehn and Dietmar Schmidt thank Cornelia Glaesser, Susan S. Kropschot, Clark Newcomb, Heinz Rosemann and Henry Toms for their examination of the manuscript as well as numerous useful hints on its revision. Hans-Georg Carls (Aerial Photo Databank/Luftbilddatenbank in Würzburg) is thanked for his advice on wartime aerial photos and archival documents. Thanks also to our coworkers of BGR Department B1.17, led by Dieter Bannert, for their help in completing the extensive methodological and experimental investigations.

It should be emphasized that it was due to the constructive help and kindness of the Märkischen Entsorgungsanlagen-Betriebsgesellschaft mbH (MEAB) company that we were able to carry out the multitude of tests, experiments, and random sampling at the landfill site.

The research of John M. Irvine and coauthors was supported in part by the Strategic Environmental Research and Development Program (SERDP) and by the Environmental Program's Government Applications Task Force (GATF). The project was conducted jointly by the Environmental Research Institute of Michigan (ERIM) and the Oak Ridge National Laboratory (ORNL). ORNL participants included the ER Remote Sensing Program, the ER WAG 4 Site Investigation EPIC team, and the National Security Program Office. The authors wish to thank Ms. Susan Edginton for graphics and production assistance.

Vernon Singhroy would like to thank Robert St. Jean and Jose Leveque for processing the Sudbury images, and the Environmental Division of INCO for permission to use their site for our study.

The manuscript of Trude King and coauthors has benefitted from the helpful comments and reviews of Ian Ridley and Eric Livo. The authors would like to thank Robert Green for his help in acquiring the AVIRIS data.

Douglas C. Peters and Phoebe L Hauff would like to thank Sandra Perry of Perry Remote Sensing Ltd. and William Peppin of Advancd Software Application, Inc. for providing processed satellite images of the Goldfields, Nevada area and Eric

Livo of the U.S. Geological Survey for his *AVIRIS* images of the Cripple Creek, Colorado area and his technical support and information resulting from his study of the area (Livo 1994).

Thanks also to the Cripple Creek and Victor Gold Mining Co., particularly John Hardaway and Jeff Pontius, for allowing access to their properties in the Cripple Creek area and providing additional information and support during visits by Peters and Hauff to the area. The understanding and support of the mining industry plays a vital role in successfully performing remote sensing and environmental research on old, yet still active, mining districts.

Woody Barry and James Sjoberg of the former Reno Research Center of the U.S. Bureau of Mines are thanked for carrying out ICP and SEM analyses of the samples from the Cripple Creek District. Their perseverance with this work during the last months of the Bureau shutdown is a tribute to their professionalism and greatly aided bringing this mine waste research to successful completion.

Last but not least we are indebted to Ingrid Boller, Elke Graf, Claudia Wiessner, Matthias Sack, Steffen Pruefer, and Claudia Kirsch (all at the BGR) for their careful drawings and detailed figures, and for contributing substantially to the layout and design of this handbook.

Sources of Maps, Photos, and Images

Chapter 1–4 and Sections 6.2 and 6.3

- Aerial Photo Databank (Luftbilddatenbank) in Würzburg, Saalgasse 3 u. 5, 97082 Würzburg
- Berliner Spezialflug, Luftbild GmbH, Wassmannsdorfer Str., 15831 Diepensee
- Boeker, F., Kopernikus-Straße 3, 30982 Pattensen
- City of Halle Environmental Bureau (Stadt Halle, Saale, Umweltamt), Marktplatz 1, 06108 Halle (Saale)
- Environmental and Economic Geology Co. Ltd. (Gesellschaft für Umwelt- und Wirtschaftsgeologie mbH; UWG), now: Fugro Consult GmbH, Wolfener Straße 36, 12681 Berlin
- Eurosense GmbH, Markgrafendamm 24, 10245 Berlin
- Federal Armed Forces Geographic Office (Amt für Militärisches Geowesen), Frauenberger Straße 250, 53879 Euskirchen
- Federal Institute for Geosciences and Natural Resources (Bundesanstalt für Geowissenschaften und Rohstoffe) Stilleweg 2, 30655 Hannover
- Federal Institute for Geosciences and Natural Resources, Berlin Branch (Bundesanstalt für Geowissenschaften und Rohstoffe, Dienstbereich Berlin), Wilhelmstraße 25–30, 13593 Berlin
- Federal Archive, Potsdam Branch (Bundesarchiv, Abteilung Potsdam) Berliner Straße 98–101, 14467 Potsdam
- FKP Engineering Office for Remote Sensing, Photogrammetry, and Cartography Ltd. (FPK Ingenieurbüro für Fernerkundung, Photogrammetrie und Kartographie GbR), Feurigstraße 54, 10627 Berlin
- German Aerospace Research Establishment (Deutsche Forschungsanstalt für Luft- und Raumfahrt, DLR), 82234 Oberpfaffenhofen
- Hansa Luftbild GmbH, Elbestraße 5, 48145 Münster
- KAZ Bildmess GmbH, Karl-Rothe-Straße 10–14, 04133 Leipzig
- Merkt, Dr. J., Niedersaechsisches Landesamt für Bodenforschung, Stilleweg 2, 30655, Hannover
- State Survey Office (Landesvermessungsamt) Brandenburg, Heinrich-Mann-Allee 103, 14437 Potsdam
- Senate for Building, Housing and Transport Berlin, Aerial Photo Archives (Senatsverwaltung für Bau- und Wohnungswesen Berlin, Luftbildarchiv) Mansfelder Straße 16, 10713 Berlin
- Thuringia State Administration Office (Thüringer Landesverwaltungsamt), State Survey Office (Landesvermessungsamt), Anger 6, 99084 Erfurt

- uve GmbH, Berliner Straße 50, 14467 Potsdam
- WIB Ingenieurgesellschaft GmbH, Lassenstraße 11–15, 14193 Berlin
- WTJ Software Services, 809 Lawrence Road, San Mateo, CA 94401, USA.

Official Permissions

Aerial Photographs

- Figure 6.4 (1981) reproduced with the permission of the Thüringer Landesverwaltungsamt on 16.02.1995,
- Figures 3.3, 4.2, 4.10, 6.15 (above) reproduced with authorization no. LBB XI/95 from Landesvermessungsamt Brandenburg on 31.01.1995,
- Figure 4.5 reproduced with the permission of the Berliner Senatsverwaltung für Bau- und Wohnungswesen V on 30.01.1995,
- Figures 4.3, 6.15, 6.16 reproduced with the permission of the Bundesarchiv (Federal Archives), Potsdam Branch on 20.04.1994,
- Figure 6.4 (1961, 1971) reproduced with the permission of the Bundesarchiv (Federal Archives), Potsdam Branch on 28.02.1995.

Maps

- Figures 4.3, 6.22, 6.23 reproduced with authorization no. GB-D 4/95 from the Landesvermessungsamt Brandenburg on 22.02.1995 (Top. Grundlage TK 10AV and TK25AS).

Glossary of Frequently Used Abbreviations

AMSS	Advanced multi spectral scanner
AP	Aerial photo
AVIRIS	Airborne visible/infrared imaging spectrometer
BGR	Federal Institute for Geosciences and Natural Resources (*Bundesanstalt für Geowissenschaften und Rohstoffe*)
BW	Black-and-white (photography)
CASI	Compact airborne spectrographic imager
CCD	Charge coupled device
CET	Central European Time (UTC + 1 hr)
CIR	Color infrared
c_f	Focal length (of aerial photographic camera)
CRC	Color ratio composite
DAIS	Digital airborne imaging spectrometer
DEM	Digital Elevation Model
DLR	German Aerospace Center (*Deutsches Zentrumt für Luft- und Raumfahrt e.V.*)
DOE	U.S. Department of Energy
DOI	U.S. Department of the Interior
ERIM	Environmental Research Institute of Michigan
EM	electromagnetic
ENVI	Environment for Visualizing Images
EPA	U.S. Environmental Protection Agency
FCC	False-color composite
FOV	Field of view
GER	Geophysical Environmental Research
GERIS	Geophysical Environmental Research imaging spectrometer
GIS	Geographic information system or geobased information system
GPS	Global positioning system
GSD	Ground sample distance
h_g	Flight altitude (height above ground)
IFOV	Instantaneous field of view
IFSAR	Interference Synthetic Aperture Radar
IR	Infrared
JERS-1	Japanese Earth Resources Satellite
K	Kelvin
λ	Wavelength

LAGA	State Waste Association, Germany (*Länderarbeitsgemeinschaft Abfall*)
Landsat TM	Landsat Thematic Mapper
LDK	Limonite and degraded kaolin
MEIS	Multidetector electro-optical imaging scanner
MIR	Middle infrared
MIVIS	Multispectral Infrared and Visible Imaging Spectrometer
MSS	Landsat multispectral scanner
NAPP	National Aerial Photography Program of the U.S. Geological Survey
NASA	National Aeronautics and Space Administration, U.S.
NDVI	Normalized Difference Vegetation Index)
NIR	Near infrared
NLfB	Lower Saxony Geological Survey (*Niedersächsisches Landesamt für Bodenforschung*)
nm	Nanometer
NUV	Near ultraviolet
ORNL	Oak Ridge National Laboratory, U.S.
ORR	Oak Ridge Reservation
PIMA	Field-portable short wave infra-red spectrometer
QSP	Quartz-sericite-pyrite
RTGC	Radiative transfer ground calibration
S	Siemens, unit of electrical.conductivity
SAM	Spectral angle mapper algorithm
SAR	Synthetic aperture radar
SEM	Scanning electron microscope
SWIR	Short-wave infrared
SWSA	Solid waste storage area
TK	Topographical survey sheet/map (abbreviation on German maps)
TM	Landsat Thematic Mapper
USBM	U.S. Bureau of Mines
USGS	U.S. Geological Survey
UTM	Universal Transverse Mercator
UV	Ultraviolet
VIS	Visible light
WAG	Waste Area Group
WW I	World War I
WW II	World War II

References

Adams JB (1975) Visible and near-infrared diffuse reflectance of pyroxenes as applied to remote sensing of solid objects in the solar system. J Geophys Res 79:4829–4836
Ahrens H (1992) Verbundvorhaben Deponieuntergrund: Teststandort Schöneiche/Schöneicher Plan (Vorstudie), UWG mbH Berlin im Auftrag der BGR, (unpubl.)
Albertz J (1991) Grundlagen der Interpretation von Luft- und Satellitenbildern: Eine Einführung in die Fernerkundung. Wiss. Buchges., Darmstadt
Allen HE, Perdue EM, Brown DS eds (1993) Metals in groundwater. Lewis Publishers, Boca Raton, Florida
Anger CD, Babey SK, Adamson RA (1990) A new approach to Imaging Spectrocscopy. Proceedings of SPIE 72:72–86
Ashley RP (1974) Goldfield Mining District. In: Guidebook to the geology of four Tertiary volcanic centers in central Nevada. Nevada Bureau of Mines, Report 19:49–66
Ashley RP (1977) Relation between volcanism and ore deposition at Goldfield, Nevada. In: Proc of the 5th IAGOD Quadrennial Symp, pp 77–85
Ashley RP (1979) Relations between volcanism and ore deposition at Goldfield, Nevada. Nevada Bureau of Mines and Geology Report No. 33
Ashley RP, Albers JP (1975) Distribution of gold and other ore-related elements near ore-bodies in the oxidized zone at Goldfield, Nevada. U.S. Geological Survey Professional Paper 843-A
Ashley RP, Silberman ML (1976) Direct dating of mineralization at Goldfield, Nevada by potassium-argon and fission track methods. Econom Geol 71:904–924
Azcue JM, Nriagu JO (1995) Impact of abandoned mine tailings on the arsenic concentrations in Moira lake, Ontario. Journal of Geochemical Exploration 52:81–89
Azcue JM, Murdoch A, Rosa F, Hall GEM, Jackson TA, Reynoldson T (1995) Trace elements in water, sediments, porewater, and biota polluted by tailings from an abandoned gold mine in British Columbia, Canada. J Geochem Explor 52:25–34
Babey SK, Anger CD (1989) A compact airborne spectrographic imager (CASI). IGARSS Proceedings 2:1028–1031
Barrett EC, Curtis LF (1992) Introduction to environmental remote sensing. Chapman Hall, London
Beckett PJ, Negusanti J, Peters T, Vining J, Miller J, Lautenbach W (1995) The Sudbury regional Land Reclamation Program. Tree and shrub enhancement. Mining and the environment. Proceedings Sudbury 95:1103–1112
Berger BR (1986) The geological attributes of Au-Ag-base metal epithermal deposits. In: Erickson R, comp., Characteristics of mineral-deposit occurrences: USGS Open File Report 82-795:119-126
Bianci R, Cavalli RM, Fiumi L, Maraino CM, Pignatti S (1997) Airborne remote sensing: Results of two years of imaging spectrometry for the study of environmental problems. In: Remote Sensing '96, Balkema, Rotterdam, pp 269–273
Blanchard R (1968) Interpretation of leached outcrops. Nevada Bureau of Mines, Bulletin 66, 196
Boeker F, Kuehn F (1992) Zur Verwendung von Luftbildern bei der geologischen Kartierung. Z angew Geol 38, 2:80–85

Boldt CMK, Scheibner BJ (1987) Remote sensing of mine waste. U.S. Bureau of Mines Information Circular 9152, 43
Bormann P (1981a) Passive Fernerkundungssensoren im optischen Bereich des Spektrums. Vermessungstechnik 29, 2:45–48
Bormann P (1981b) Was ist unter dem Auflösungsvermögen eines Fernerkundungssensors zu verstehen. Vermessungstechnik 29, 10:331–335
Brown RL (1991) Cripple Creek then and now. Sundance Publications, Ltd., Denver
Brückner G, Knitschke G, Spilker M, Pelzel J, Schwandt A (1983) Probleme und Erfahrungen bei der Beherrschung von Karsterscheinungen in der Umgebung stillgelegter Bergwerke des Zechsteins in der DDR. Neue Bergbautechnik, Leipzig 13, 8:417–422
Ciciarelli JA (1991) A practical guide to aerial photography. Van Nostrand, Würzburg
Clark RN, Gallagher AJ, Swayze GA (1990a) Material absorption band depth mapping of imaging spectrometer data using a complete band shape least-squares fit with library reference spectra. In: Proceedings of the second airborne visible/infrared imaging spectrometer (AVIRIS) Workshop, JPL Publication 90-54:176–186
Clark RN, King TVV, Klejwa M, Swayze G, Vergo N (1990b) High spectral resolution reflectance spectroscopy of minerals. J Geophys Res 95:12653–12680
Clark RN, Swayze GA, Gallagher A, Gorelick N, Kruse F (1991) Mapping with imaging spectrometer data using the complete band shape least-squares algorithm simultaneously fit to multiple spectral features from multiple materials. In: Proc of Third Airborne Visible/Infrared Imaging Spectrometer (AVIRIS) Workshop in Pasadena, CA, on May 20–21 1991, JPL Publication 91-28:2–3
Clark RN, Swayze GA, Koch C, Gallagher A, Ager C (1992) Mapping vegetation types with the multiple spectral feature mapping algorithm in both emission and absorption. In: Summaries of the Third Annual JPL Airborne Geosciences Workshop, vol. 1: AVIRIS Workshop, held in Pasadena, CA, on June 1–5, 1992. JPL Publication 92-14:60–62
Clark RN, Swayze GA, Gallagher A, King TVV, Calvin WM (1993) The U. S. Geological Survey, Digital Spectral Library: Version 1: 0.2 to 3.0 μm, USGS, Open File Report 93-592, 1340 pp (also being published as a USGS Bulletin, 1300 + pp, 1996 in press.)
Clark RN, King TVV, Ager C, Swayze GA (1995) Initial vegetation species and sene-scence/stress mapping in the San Luis Valley, Colorado using imaging spectrometer data. In: Summaries of the Fifth Annual JPL Airborne Earth Science Workshop, January 23–26, R.O. Green, Ed., JPL Publication 95-1:35–38
Cloutis EA, Gaffey MJ (1987) Pyroxene spectroscopy revisited: Spectral-compositional correlations and relationships to geoherometry. J Geophys Res 96:22809–22826
Cloutis EA, Gaffey MJ, Jackowski TL, Reed KL (1986) Calibrations of phase abundance, composition, and particle size distribution of olivine-orthopyroxene mixtures from reflectance spectra. J Geophys Res 91:11641–11653
Colwell RN (ed) (1983) Manual of remote sensing, vol. 1: Theory, instruments and techniques, vol. 2: Interpretation and applications. American Society of Photogrammetry, Falls Curch
Cox DP, Bagby WC (1986) Model 22b – Descriptive model of Au-Ag-Te veins. In: Cox DP, Singer DA (eds) (1986) Mineral deposit models. USGS Bulletin 1693, 124
Cox DP, Singer DA (eds)(1986) Mineral deposit models. USGS Bulletin 1693, 379
Crawford GA (1995) Environmental improvements by the mining industry in the Sudbury Basin of Canada. J Geochem Explor 52:267–284
Curran PJ (1989) Remote sensing of foliar chemistry. Remote Sens Environ 30:271–278
Curran PJ, Dungan JL, Macler BA, Plummer SE (1991) The effect of a red leaf pigment on the relationship between red edge and chlorphyll concentration. Remote Sens Environ 35:69–76
Curran PJ, Dungan JL, Macler BA, Plummer SE, Peterson DL (1992) Reflectance spectroscopy of fresh whole leaves for the estimation of chemical concentration. Remote Sens Environ 39:153–166
Danks SM, Evans EH, Whittacker PA (1984) Photosynthetic systems: Structure, function and assembly. Wiley, New York

References

Davis AD, Webb CJ (1995) Abandoned mines inventory and reclamation in the Black Hills of South Dakota. In: Scheiner BJ, Chatwin TD, El-Shall H, Kawatra SK, Torma AE (eds) New remediation technology in the changing environmental arena. Littleton, Colorado. Society for Mining, Metallurgy, and Exploration, Inc., pp 27–33

Dech SW, Glaser R, Kuehn F, Carls H-G (1991) Ökologische Probleme durch Rüstungsaltlasten in der Colbitz-Letzlinger Heide. DLR-Nachrichten, Heft 64

Despain, DG (1990) Yellowstone vegetation, consequences of environment and history in a natural setting, Roberts Rinehart Publishing

De Vos KJ, Blowes DW, Robertson WD, Greenhouse JP (1995) Delineation and evaluation of a plume of tailings derived water, Copper Cliff, Ontario. Mining and the Environment, Proceedings, Sudbury, pp 673–682

Dodt J, Borries HW, Echterhoff-Friebe M, Reimers M (1987) Zur Verwendung von Luftbildern und Karten bei der Ermittlung von Altlasten. Ruhr-Universität Bochum, Anlagenband

Dorn RI, Oberlander TM (1981) Microbial origin of desert varnish. Science 213:1245–1247

Douglas WJ (1995) Environmental GIS – Applications to industrial facilities. Lewis Publishers, Boca Raton, FL

Durkin ThV (1995) Sulfide mine waste management: the need for new technology development. The Professional Geologist 33, 2:4–7

Ehrenberg M (1991) Beprobungslose Altlastenerkundung. wlb Wasser, Luft und Boden 7–8:56–59

Erb W (1989) Leitfaden der Spektroradiometrie. Springer, Heidelberg Berlin New York, 386 pp

Fernandez HM, Veiga LHS, Franklin MR, Prado VCS, Taddei JF (1995) Environmental impact assessment of uranium mining and milling facilities: A case study at the Poros de Caldas uranium mining and milling site, Brazil. J Geochem Explor 52:161–173

Gaffey SJ (1986) Spectral reflectance of carbonate minerals in the visible and near-infrared (0.35–2.55 microns); Calcite, aragonite, and dolomite. Am Mineral 71:151–162

Gaffey SJ (1987) Spectral reflectance of carbonate minerals in the visible and near-infrared (0.35–2.55 µm) Anhydrous carbonate minerals. J Geophys Res 92:1429–1440

Gaffey SJ, McFadden LA, Nash D, Pieters CM (1993) Ultraviolet, visible, and near-infrared reflectance spectroscopy: Laboratory spectra of geologic materials. In: Pieters CM, Englert PAJ (eds) Remote chemical analysis: Elemental and mineralogical composition. Cambridge University Press, pp 43–77

Gebhardt A (1981) Thermografie, Anwendungen bei der geophysikalischen Naherkundung. Freiberger Forschungsheft, C367, 16 Beilagen, Deutscher Verlag für Grundstoffindustrie, Leipzig

Glass CE, Schowengerdt RA (1983) Hazard and risk mapping of mined lands using satellite imagery and collateral data. Bull Assoc Eng Geol 20, 2:205–218

Gleason VE, Russell HH comps. (1995) Coal and the environment abstract series. Mine drainage bibliography 1910–1976: Monroeville, Pennsylvania, Bituminous Coal Research, Inc.

Gonzalez H, Ramirez M (1995) The effect of nickel and metallurgical activities on the distribution of heavy metals in Levisa Bay, Cuba. J Geochem Explor 52:183–192

Goodchild MF, Steyaert LT, Parks BO, Johnston C, Maidment D, Crane M, Glendinning S (eds) (1996) GIS and envronmental modeling. Progress and research issues: GIS World Books, Fort Collins, Colorado

Gott GB, McCarthy JH Jr., VanSickle GH, McHugh JB (1969) Distribution of gold and other metals in the Cripple Creek District, Colorado. USGS Professional Paper 625-A

Graham DF, St-Arnand EL, Rencz AN (1994) Canada Geologic Survey Monitors Mine Tailings, Disposal sites with Landsat. EOS Magazine, pp 38–41

Grimstad WN, Drake RL (1983) The last gold rush: A pictorial history of the Cripple Creek and Victor gold mining district. Pollux Press, Victor, Colorado

Gundermann DG, Hutchinson TC (1995) Changes in soil chemistry 20 years after the closure of a nickel-copper smelter near Sudbury, Ontario, Canada. J Geochem Explor 52:231–236

Gunn JM (ed) (1995) Restoration and recovery of an industrial region. Springer, Heidelberg Berlin New York

Gupta RP (1991) Remote Sensing Geology. Springer, Heidelberg Berlin New York

Harvey RD, Vitaliano CJ (1964) Wall rock alteration in the Goldfield District, Nevada. J Geol 72:564–579

Hauff Ph L, (1993) Spectral reflectance properties of oil and hydrocarbon-bearing rocks and sediments. Application Note, No. 1, 1993, Spectral International Inc., Lafayette, CO

Hornsby JK, Bruce B, Mackenzie-Greive G (1989) Monitoring vegetation regrowth on placer mine tailings, Bonanza Creek, Yukon Territory. In: Proc of Int Symp on Remote Sensing, pp 2518–2521

Horowitz AJ, Elrick KA, Robbins JA, Cook RB (1995) A summary of the effects of mining and related activities on the sediment-trace element geochemistry of Lake Cour d'Alene, Idaho, USA. J Geochem Explor 52:135–144

Hunt GR (1977) Spectral signature of particulate minerals in the visible and near infrared. Geophysics 42:501–513

Jansen WT (1994) The mapping of mineral distributions using remotely sensed hyperspectral images and standard spectral libraries. Int. Symposium on Remote Sensing and GIS for Site Characterizations – Applications and Standards, ASTM, San Francisco, Jan. 27–28 1994

Johns C (1995) Contamination of riparian wetlands from past copper mining and smelting in the headwaters region of the Clark Fork River, Montana, U.S.A.. J Geochem Explor 52:193–203

Johnson AI, Pettersson CB, Fulton JL (eds) (1992) Geographic information systems (GIS) and mapping – Practices and standards: ASTM, Publication STP 1126

King DJ (1993) Digital frame cameras: the next generation of low cost remote sensing sensors. In: Proc of ASPRS Biennial Workshop on Photography and Videography in the Plant Sciences, Logan, Utah

King TVV (ed) (1995) Environmental considerations of active and abandoned mine lands: Lessons from Summitville, Colorado. USGS Bulletin 2220

King TVV, Clark RN (1989) Spectral characteristics of chlorites and Mg-serpentine using high-resolution reflectance spectroscopy. J Geophys Res 94:13997–14008

King TVV, Ridley WI (1987) Relation of the spectroscopic reflectance of olivine to mineral chemistry and some remote sensing implications. J Geophys Res 92:11457–11469

Kokaly RF, Clark RN, Livo KE (1998) Mapping the biology and mineralogy of Yellowstone National Park using imaging spectroscopy. In: Summaries of 7th Annual JPL Airborne Earth Science Workshop, R.O. Green, ED., JPL Pub 97-21, vol 1

Koschmann AH (1949) Structural control of the gold deposits of the Cripple Creek district, Teller County, Colorado. USGS Bulletin 955-B, 60

Krenz O (1991) Luftbildinterpretation der Deponiestandorte Schöneiche und Schöneicher Plan. Teilbericht zum BMFT-Projekt „Abfallwirtschaftliche Rekonstruktion von Altdeponien am Beispiel des Deponiestandortes Schöneiche", (unpubl.)

Kronberg P (1984) Photogeologie, eine Einführung in die Grundlagen und Methoden der geologischen Auswertung von Luftbildern. Enke, Stuttgart

Kronberg P (1985) Fernerkundung der Erde. Enke, Stuttgart

Kruse FA, Hauff PL, Dietz J, Brock JC, Hampton LR (1989) Characterization and mapping of mine waste at Leadville, Colorado using imaging spectroscopy. Boulder, Center for the Study of Earth from Space, U of Colorado, EPA Contract No. 68-01-7251, 2 vols

Kuehn F (1992) Ergebnisse der Luftbildauswertung für den Deponiestandort Eulenberg bei Arnstadt; Teil I: Auswertung und Interpretation historischer Luftbilder; Teil II: Photogeologische Beurteilung des Deponiestandortes. Bundesanstalt für Geowissenschaften und Rohstoffe, Außenstelle Berlin, (unpubl.)

Kuehn F, Hoerig B (1995) Handbuch zur Erkundung des Untergrundes von Deponien und Altlasten, Band 1: Geofernerkundung. Springer, Berlin Heidelberg New York

Kuehn F, Oleikiewitz P (1983) Die Nutzung der Multispektraltechnik zur Früherkennung von senkungs- und erdfallgefährdeten Gebieten. Z angew Geol 29, 6:71–74

Kuehn F, Knoedel K, Krummel H, Lange G (1994) Kombinierte Nutzung von Luftbildern und geophysikalischen Methoden bei der Untersuchung eines Deponiestandortes. Z angew Geol 40, 2:61–68

Kuehn F, Trembich G Hoerig B (1997) Multisensor remote sensing to evaluate hazards caused by mining. In: Proceedings of the Twelfth International Conference on Applied Geologic Remote Sensing, 17–19 November, Denver, Colorado, ERIM, I:425–432

Kuehn F, Trembich G, Hoerig B (1999) Satellite and airborne remote sensing to detect hazards caused by underground mining. In: Proceedings of the Thirteenth International Conference on Applied Geologic Remote Sensing, 1–3 March, Vancouver, British Columbia, Canada, ERIM, II:57–64

Lee K (1989) Limonite mapping with Landsat multispectral scanner data at Cripple Creek, Colorado. In: Keenan Lee (ed) Remote sensing in exploration geology: 28th International Geological Congress Field Trip Guidebook T-182:8–12

Lee MB (1958) Cripple Creek days: New York, Doubleday, Co., Inc.

Lindgren W, Ransome FL (1906) Geology and gold deposits of the Cripple Creek District, Colorado. USGS Professional Paper 54

Livo KE (1994) Use of remote sensing to characterize hydrothermal alteration of the Cripple Creek area, Colorado. Golden, Colorado School of Mines, M. Sc. Thesis T-4613, (unpubl.)

Loeffler J (1962) Die Kali- und Steinsalzlagerstätten des Zechsteins in der Deutschen Demokratischen Republik. Freiberger Forschungsheft, Teil III Sachsen-Anhalt, C97/III, 16 Beilagen, Akademie Verlag, Berlin

Lowe DS (1969) Optical sensors. In: Principles and applications to earth resources surveys. CNS and Univ. of Michigan, Paris, pp 73–136

Lyon JG, McCarthy J (eds) (1995) Wetland and environmental applications of GIS. Lewis Publishers, Boca Raton, Florida

Lyon JS, Hilliard TJ, Bethell TN (1993) Burden of gilt. Mineral Policy Center, Washington, DC

Lyon RJP (1994) Weathering and desert varnish in arid terrains. In: Proceedings of the First International Airborne Remote Sensing Conference and Exhibition, Strasbourg, France, Sept. 11–15, 1994. Ann Arbor, Environmental Research Institute of Michigan 1:257–268

Mausel PW, Howe RC, Lulla K (1981) Non-parametric classification of abandoned coal mine features using multioriented and ratio transformed Landsat data. In: Proc of 15th Int Symp on Remote Sensing of Environment, Ann Arbor, Michigan, May 1981: Ann Arbor, Environmental Research Institute of Michigan, pp 1397–1409

Maxwell EL (1974) Automatic mapping of strip and open pit mine excavations – using ERTS imagery. Fort Collins, Colorado, Marlatt and Associates, U.S. Bureau of Mines Contract No. SO241090, 59

McCann ME, King A (1995) Case study of advanced technologies for hydrological site characterization. Mining and the Environment, Sudbury 95 Proceedings, Sudbury, pp 667–671

McGregor RG, Blowes DW, Robertson WD (1995) The Application of chemical extractions to sulfide tailings at the copper cliff tailings area, Sudbury, Ontario. Mining and the Environment, Sudbury 95, Proceedings, Sudbury, pp 1133–1142

Merkt J, Boeker F (1993) Erkundung von quartärgeologischen Bildungen mit saisonalen Luftbildern. Geol Jb A142:65–88

Meyer W (1993) Abschlußbericht Reflexionsseismik Eulenberg/Arnstadt. Geophysik GGD Leipzig, (unpubl.)

Miller JR, Hare EW, Hollinger AB, Sturgeon DR (1987) Imaging spectroscopy as a tool for botanical mapping. In: Vane G (ed) Imaging spectroscopy II. International Society for Optical Engineering, Bellingham, WA, pp 108–113

Milton NM, Collins W, Chang SH, Schmidt RG (1983) Remote detection of metal anomalies on Pilot Mountain, Randolf County, North Carolina. Econ Geol 78:605–617

Milton NM, Ager CM, Eiswerth BA, Power MS (1989) Arsenic and selenium-induced changes in spectral reflectance and morphology of soybean plants. Remote Sens Environ 30:263–269

Mined Land Reclamation Bureau (1982) Their silent profile: inactive coal and metal mines of Colorado. Colorado Department of Natural Resources, Colorado Inactive Mine Reclamation Plan, vol. I, Chap II

Mining and Minerals Branch (1994) Abandoned mineral lands in the national parks. Land Resources Division, National Park Service, Informational Brochure D-904

Munts SR, Hauff PL, Seelos A, McDonald B (1993) Reflectance spectroscopy of selected base-metal bearing tailings with implications for remote sensing. In: Proc of 9th Thematic Conf on Geol Remote Sensing, Pasadena, CA, Feb. 8–11, 1993: Ann Arbor, Environmental Research Institute of Michigan, pp 567–578

Murray I, Williams PC (1987) Chemical principles of near-infrared. In: Williams P, Norries K (eds) Near infrared technology in the agricultural and food industries. American Association of Cereal Chemists, St. Paul, MN, pp 17–37

Mussokowski R (1983) A technique for mapping environmental change – using digital Landsat data. COSPAR, Advances in Space Research 8:103–107

Mussokowski R, Chan P, Goba N (1993) Spatial modelling of abandoned mine tailings for environmental assessment – The Kam Kotia test site. Ontario Ministry of Natural Resources Geographic Information Seminar

Negusanti JJ (1995) Terrestrial metal trends in the Sudbury Area. Mining and the environment. Sudbury 95 Proceedings, Sudbury, pp 1143–1149

Parsiegla, K (2000). Handbuch zur Erkundung des Untergrundes von Deponien und Altlasten, Band 8: Fallstudien, Springer, Berlin Heidelberg New York, in prep.

Plumlee GS, Smith KS, Ficklin WH, Briggs PH, McHugh JB (1993) Empirical studies of diverse mine drainages in Colorado: Implications for the prediction of mine-drainage chemistry. In: Planning, rehabilitation and treatment of disturbed lands. Sixth Billings Symposium, Billings, MT, March 21–27, 1993, Reclamation Research Unit, Montana State University, Publication No. 9301, 1:176–186

Post JL, Noble PN (1993) The near-infrared combination band frequencies of dioctahedral smectites, micas, and illites. Clay and Clay Minerals 41/6:639–644

Potter RM, Rossman GR (1979) The manganese- and iron-oxide mineralogy of desert varnish. Chemical Geology 25:79–94

Puro MJ, Kipkie WB, Knapp RA, McDonald TJ, Stuparyk RA (1995) Inco copper cliff tailings area. Mining and the environment, Sudbury 95 Proceedings, Sudbury, pp 181–191

Ransome FL (1909) Geology and ore deposits of Goldfield, Nevada. USGS Professional Paper 66

Renton JJ, Hidalgo RV, Streib DL (1973) Relative acid-producing potential of coal. West Virginia Geological and Economic Survey, Environmental Geology Bulletin 11, 8

Rock BA, Vogelmann DJ, Williams DL, Vogelmann AF, Hoshizaki T (1985) Remote detection of forest damage. Bio Sci 36:439

Rowan JS, Barnes SJA, Hetherington SL, Lambers B, Parsons F (1995) Geomorphology and pollution – the environmental impacts of lead mining, Leadhills, Scotland. J Geochem Explor 52:57–65

Sabins FF (1996) Remote sensing – Principles and interpretation, 3rd Edition. Freeman, San Francisco

Salomons W (1995) Environmental impact of metals derived from mining activities. Processes, predictions, prevention. J Geochem Explor 52:5–23

Schneider S (1974) Luftbild und Luftbildinterpretation. de Gruyter, Berlin

Schulze E, Seidel K, Seidemann O (1992) Ergebnisbericht über magnetische, gravimetrische, geoelektrische und refraktionsseismische Messungen am Teststandort Eulenberg. Geophysik GGD Leipzig, (unpubl.)

Sengupta M (1993) Environmental impacts of mining. Lewis Publishers, Boca Raton, Florida

Shilin BV (1980) Thermal aerial survey in natural resources research (Russian). Gidrometeoizdat, St. Petersburg

Schuchman RA, Davis CP, Jackson PL (1975) Contour strip-mine detection and identification with imaging radar. Bull Assoc Eng Geol 12, 2:99–118

Singhroy VH (1995) Spectral characterization of vegetation at mine tailings. Mining and the Environment, Sudbury 95 Proceedings, Sudbury, pp 193–200

Singhroy VH, Kruse F (1991) Detection of metal stress in the boreal forest species using the 670 µm chlorophyll band. ERIM, 8th Thematic Conf on Geologic Remote Sensing, pp 361–372

References

Singhroy VH, Kenny F, Springer J (1989) Reflectance spectra of vegetation growing on mine site in the Canadian Shield. In: 12th Canadian Symp on Remote Sensing, Proceedings. IGARSS '89, Vancouver, B.C., Canada, pp 665–669

Stewart KC, Severson RC (eds) (1994) Guidebook on the geology, history, and surface-water contamination and remediation in the area from Denver to Idaho Springs, Colorado. USGS Circular 1097, 55

Strathmann F-W (1993) Taschenbuch zur Fernerkundung. Wichmann, Karlsruhe

Struhsacker DW (1995) The importance of waste characterization in effective environmental planning, project design and reclamation. In: Scheiner BJ, Chatwin TD, El-Shall H, Kawatra SK, Torma AE (eds) New remediation technology in the changing environmental arena. Society for Mining, Metallurgy, and Exploration, Inc., Littleton, Colorado, pp 19–25

Swayze GA, Clark RN (1990) Infrared spectra and crystal chemistry of scapolites: implications for Martian mineralogy. J Geophys Res 95:14481–14495

Taranik DL (1990) Remote detection and mapping of supergene iron oxides in the Cripple Creek mining district, Colorado. U. of Colorado, M.Sc. Thesis, (unpubl.)

Taylor HP Jr. (1973) O18/O16 evidence for meteoric-hydrothermal alteration and ore deposition in the Tonopah, Comstock Lode and Goldfield mining districts, Nevada. Econom Geol 68, 6:747–764

Theilen-Willige, R. (1993) Umweltbeobachtung durch Fernerkundung. Enke, Stuttgart

Thompson TB (ed) (1986) Cripple Creek mining district. Denver Region Exploration Geologists Society, Fall Field Guidebook

Thompson TB, Trippel AD, Dwelley PC (1985) Mineralized veins and brecias of the Cripple Creek District, Colorado. Econom Geol 80:1669–1688

Trevett JW (1983) Imaging radar for resource surveys. Chapman & Hall, London

U.S. Forest Service (1993) Acid drainage from mines on the national forests: a management challenge. Program Aid 1505, 12

Vane G et al. (1993) The airborne visible/infrared imaging spectrometer (AVIRIS). Remote Sensing Environ 44:127–143

Vangronsveld J, Sterckx J, Van Assche F, Clijsters H (1995) Rehabilitation studies on an old non-ferrous waste dumping ground. Effects of revegetation and metal immobilization by beringite. J Geochem Explor 52:221–229

Winterhalder K (1984) Environmental degradation and rehabilitation in the Sudbury area. Laurentian University Rev 16, 2:15–47

Wooding MG (1988) Imaging radar applications in Europe, illustrated experimental results (1978–1987). ESA TM-01, Noordwijk, The Netherlands

Wunderlich J (1991) Vorstudie zum Forschungsverbundvorhaben Deponieuntergrund, Untersuchungsobjekt: Deponie am Eulenberg westlich Arnstadt/Thüringen. GEOS Ingenieurbüro Jena GmbH, (unpubl.)

Wunderlich J (1992) Ergebnisbericht über die Resultate der Schürfe an der ehemaligen Deponie Eulenberg in Landkreis Arnstadt. GEOS Ingenieurbüro Jena GmbH, (unpubl.)

Zilioli E, Gomarasca MA, Tomasoni R (1992) Application of terrestrial thermography to the detection of waste-disposal sites. Remote Sensing Environ 40, 2:153–160

Additional References

AFL (Arbeitsgruppe Forstlicher Luftbildinterpreten (1988) Auswertung von Color-Infrarot-Luftbildern. Selbstverlag der Arbeitsgruppe, Druck Forstliche Bundesversuchsanstalt, Wien

AG Bodenkunde (1982) Bodenkundliche Kartieranleitung, vol 3. Schweizerbart'sche Verlagsbuchhandlung, Hannover

Ashley RP, Keith WJ (1978) Goldfield mining district, Esmeralda County, Nevada. J Geochem Explor 9/10:204–208

Bartsch B, Glowinski B, Gorgas U, Irrek J, Schulz C (1993) Bewertung und Gefährdungsabschätzung der Deponie Hermsdorf mit Fernerkundungsmethoden. Abschlußbericht für die Bundesanstalt für Geowissenschaften und Rohstoffe als Teil des Verbundprojektes „Methoden zur Erkundung und Beschreibung des Untergrundes von Deponien und Altlasten", WIB GmbH Berlin, Anlagen

Bauer HJ, Haas D (1992) Die Erfassung von altlastverdächtigen Altablagerungen und Altstandorten aufgezeigt am Beispiel des Landes Nordrhein-Westfalen. Müllhandbuch, vol. 3, Erich Schmidt, Kennzahl 4310, Lieferung 5/92, Berlin

Borries H-W (1992) Altlastenerfassung und -erstbewertung durch multitemporale Karten- und Luftbildauswertung. Vogel, Würzburg

Borries H-W, Hüttl H (1991) Beprobungsfreie Erfassung und Erstbewertung von Rüstungsaltlasten bei Verdachtsstandorten. In: Thomé-Kozmiensky (eds) Untersuchung von Rüstungsaltlasten, EF-Verlag für Energie und Umwelttechnik GmbH, Berlin

Bremer et al. (1992) Leitfaden zum Altlastenprogramm des Landes Sachsen-Anhalt. Teil 1: Erfassung und Erstbewertung der Altlastverdachtsflächen. Landesamt für Umweltschutz Sachsen-Anhalt, Halle

Burkhardt D (1991) Erfassung von Rüstungsaltlasten bei Verdachtsstandorten. In: Thomé-Kozmiensky (ed) Untersuchung von Rüstungsaltlasten, EF-Verlag für Energie und Umwelttechnik GmbH Berlin

Chang SH, Collins W (1983) Confirmation of the airborne biophysical mineral exploration technique using laboratory methods. Econom Geol 78:723–736

Cloutis EA (1989) Spectral reflectance properties of hydrocarbons: Remote sensing implications. Science 245:165–168

Collins WE (1978) Remote sensing of crop type and maturity. Photogrammetric Engineering and Remote Sensing 44/1:43–55

Collins W, Chang SH, Raines G, Canney F, Ashley R (1983) Airborne biophysical mapping of hidden mineral deposits. Econom Geol 78:737–749

Crawford MF (1987) Preliminary evaluation of remote sensing data for detection of vegetation stress related to hydrocarbon microseepage: Mist Gas Field Oregon. Proc of 5th Thematic Conf on Remote Sensing for Exploration Geology, Environmental Research Institute of Michigan 1:161–177

Curran PJ, Kupiec JA (1995) Imaging spectrometry: A new tool for ecology. In: Danson FM, Plummer SE (eds) Advances in environmental remote sensing. Wiley, Chichester, pp 71–88

Curtiss B, Ustin SL (1989) Parameters effecting reflectance of coniferous forests in the region of chlorophyll absorption. In: Proc of IGARSS '89, 12th Can Symp on Remote Sensing, Vancouver

Daniel B et al. (1990) Altlastenanalytik: Parameterliste zur branchenspezifischen Auswahl von Analysenparametern für Altstandorte. ecomed, Landsberg/Lech

DOE (Department of Energy) (1994) Waste area group 4 site investigation sampling and analysis plan. Oak Ridge National Laboratory, Oak Ridge, Tennessee. DOE/OR/01-1337 D1. Prepared by CDM Federal Programs Corporation, Oak Ridge, TN, December

Energy Systems (Martin Marietta Energy Systems, Inc.) (1994) Preliminary report for waste area group 4 at Oak Ridge National Laboratory. Oak Ridge, Tennessee. ORNL/ER-271. Oak Ridge National Laboratory, Oak Ridge, TN, October

Energy Systems (Martin Marietta Energy Systems, Inc.) (1995) Site investigation report for waste area group 4 at Oak Ridge National Laboratory. vol I. Text. ORNL/ER-279/V1. Oak Ridge National Laboratory, Oak Ridge, TN, August

Gates DM, Keegan HJ, Schleter JD, Weidner VR (1965) Spectral properties of plants. Applied Optics 4/1:11–20

Glaeser C (1989) Beiträge zur Anwendung der Multispektraltechnik für die Lösung geowissenschaftlicher Aufgaben. Dissertation B, Textband, Martin-Luther-Universität Halle-Wittenberg

Glaser R, Carls H-G (1990) Kriegsluftbilder 1940–1945: Ein Hilfsmittel bei der Verdachtsflächenermittlung von Kriegsaltlasten und in der Umweltplanung. Laufener Seminarbeitr., Laufen/Salzach, 90.1:65–73

Goetz AF, Barret NR, Rowan LC (1983) Remote sensing for exploration: An overview. Econom Geol 78/4:573–592

Haas R (1992) Konzepte zur Untersuchung von Rüstungsaltlasten. Erich Schmidt, Berlin

Holzförster B, Tiedemann M (1991) Die Bedeutung der Luftbildauswertung für die Erfassung von Rüstungsaltlasten am Beispiel Niedersachsen. In: Thomé-Kozmiensky (ed) Untersuchung von Rüstungsaltlasten. EF-Verlag für Energie und Umwelttechnik, Berlin

Hoerig B, Katzkow N, Krull P, Kuehn F, Stojanow Tz, Stojanowa V, Ulbricht G (1985) Methodische Probleme spektrometrischer Messungen an geologischen Objekten. Z angew Geol 31/8:207–212

Horler DNH, Dockray M, Barber J (1983) The red edge of plant reflectance. Int J Remote Sensing 4:273–288

Horler DNH, Barber J, Barringer AR (1980) Effects of heavy metals on the absorbance and reflectance spectra of plants. Int J Remote Sensing 1:121–136

Huber E, Volk P (1986) Deponie- und Altlastenerkundung mit Hilfe von Fernerkundungsmethoden. Wasser und Boden 1986/10:509–515

Hunt GR, Salisbury JW (1976) Visible and near-infra red spectra of minerals and rocks: XII. Sedimentary Rocks. Mod Geol 5:211–217

Kinner U, Koetter L, Niclaus M (1986) Branchentypische Inventarisierung von Bodenkontaminationen – Ein erster Schritt zur Gefährdungsabschätzung für ehemalige Betriebsgelände. Forschungsbericht 10703001, Umweltbundesamt, Berlin

Kneib WD (1990) Abfalldeponien. In: Blume H-P (1990) Handbuch des Bodenschutzes. ecomed, Landsberg, pp 412–420

Krischok-Peppernick A (1990) Historische Erkundung. In: Collins HJ, Wolf J (eds) Erfassung von Rüstungsaltlasten, TU Braunschweig, H5, Braunschweig

Kruse FA, Seznec O, Krotkov PM (1990) An expert system for geologic mapping with imaging spectrometers. In: Applications of artificial intelligence VIII. Society of Photo-Optical Instrumentation Engineers (SPIE), Proceedings 1293:904–917

LAGA (Länderarbeitsgemeinschaft Abfall) (1990) Informationsschrift Altablagerungen und Altlasten. Müllhandbuch, Bd. 3, Kennzahl 4470, Lieferung 6/90, Erich Schmidt, Berlin

LAGA (Länderarbeitsgemeinschaft Abfall) (1991) Informationsschrift Abfallarten, Stand 1990. Müllhandbuch, Bd. 2, Kennzahl 1110, Lieferung 4/91, Erich Schmidt, Berlin

Peters DC, Hauff PL, Livo KE (1995) Remote sensing for mine waste discrimination and characterization. In: Curran PJ, Robertson YC, comps., RSS 95: Remote sensing in action. Sept. 12–15, 1995, The Remote Sensing Society, Southampton, UK, pp 866–877

Preuss J, Haas R (1987) Die Standorte der Pulver-, Sprengstoff-, Kampf- und Nebelstofferzeugung im ehemaligen Deutschen Reich. Geogr Rdsch 39.10:578–584

Rat der Sachverständigen für Umweltfragen: Sondergutachten „Altlasten". Bundestags-Drucksache 11/6191 vom 03.01.1990

Rock BA (1988) Comparison of in-situ and airborne spectral measurements of the blue shift associated with forest decline. Remote Sensing Environ 24:109–127

Schneider S (1984) Angewandte Fernerkundung: Methoden und Beispiele. Vincentz, Hannover

Singhroy VH (1992) Spectral characterization of the boreal forest species associated with geochemical anomalies. International Geological Congress, Abstracts, Kyoto, Japan, 3: 980

Singhroy VH, Stanton-Gray R, Springer J (1986) Spectral geobotanical investigations of mineralized till sites. In: 5th Int Symp on Remote Sensing of Environ, Thematic Conf on Remote Sensing for Exploration Geology, Proceedings. Environmental Research Institute of Michigan, Ann Arbor, pp 523–543

Straehle KG (1985) Historisches Luftbildarchiv seit 1919, Luftbildverzeichnis. Strähle KG Luftbild, Firmenkatalog, Schondorf, Jan. 1, 1985

Tandy BC, Amos E (1985) Airborne thermal infrared linescan in geology. Proc of Intern Symp on Remote Sensing of Environment, 4th Thematic Conf "Remote Sensing for Exploration Geology", San Francisco, California, 1.–4. April

Ustin SL, Martens SN, Curtiss B, Vanderbilt VC (1994) Use of high spectral resolution sensors to detect air pollution injury in conifer forests. In: Fensternmaker LA (ed) Remote sensing applications for acid deposition. EPA publication CR81400201, pp 72–85

Voigt H et al. (1987) Hydrogeologisches Kartenwerk der DDR 1:50 000, Karte der Grundwassergefährdung. Hrsg: Zentrales Geologisches Institut und Kombinat Geologische Forschung und Erkundung, Berlin/Halle

Weber H (ed) (1993) Altlasten: Erkennen, Bewerten, Sanieren. Springer, Heidelberg Berlin New York

Subject Index

A

abandoned mine lands 113
abandoned landfill 37, 88, 89, 95
acid drainage 106–113, 143, 145, 165, 175
acid generating potential 137, 146, 150, 157
acid generator 159
acid water 113, 116, 157
acidic water 165
aerial photographs 11–20
 –, archival 17–20, 44, 57, 83
 –, black-and-white 14
 –, chlorophyll 91
 –, color 15–17, 47, 54, 74
 –, color-infrared (CIR) 15–16, 42, 43, 45, 52, 53, 65, 76, 90, 123
 –, forward overlap 12
 –, historic 17–20, 100, 101
 –, NAPP 123, 124
 –, oblique 16, 17, 47, 50, 74, 97
 –, photogrammetric interpretation 33
 –, side lap 12
 –, scale 13
 –, stereoscopic 13–14, 100
 –, thematic interpretation 33
 –, wartime 19–20, 71
airborne imaging spectrometers 28
airborne multispectral 106, 113
airborne spectroscopy 151
airborne scanners 24
airphoto camera 12
airphoto film 14–17
 –, color 15
 –, color-infrared 15
 –, infrared 14
 –, panchromatic 14
 –, resolution 13
Alamosa River 164–175, 184
anthropogenic features 34
atmosphere 7–8

B

bioindicator 91
blackbody 5–8
 –, temperature 7
bomb crater 19, 20, 45
bunker 64–73
buried waste 96–105

C

CCD array 26, 27
chlorophyll 15, 91, 175, 176, 178
 –, absorption 179, 180, 182, 184
 –, continuum-removed absorption spectra 178
CIR aerial photograph (see *color-infrared photograph*)
classification 60
color composite 109, 112
color ratio-composite (CRC) 125–131, 152, 153
contamination 14, 33, 49, 64, 87, 89, 93, 184
continuum 167
continuum removal process 167
continuum removed spectra 178
Cripple Creek mining district 116–145
 –, geology 117–123
 –, mining history 117–123

D

Daedalus scanner 23–24, 99, 101
desert varnish 144
drainage system 46–48, 83

E

electromagnetic radiation 5–8
 –, absorption 6
 –, emission 8
 –, interaction 5
 –, reflection 6
electromagnetic spectrum 6
 –, far infrared 6
 –, microwave 6
 –, middle infrared 6
 –, near ultraviolet 6
 –, visible light 6
 –, energy spectrum 6
ERS-1/2 10
Eulenberg waste disposal site
 –, aerial photographs 68
 –, geophysical investigation 66–68
 –, interpretation 72
 –, landfill 64

F

false-color composite (FCC) 124–131, 151, 152
fault 49, 50, 67
field of view (FOV) 22
field check 5, 59, 117, 152, 169
field sampling 61
fractures 49, 50
fracturing 145
frost polygon 47

G

GERIS spectrometer 123
GIS technology 162, 163
Goldfield Mining District 146–164
 –, geological background 148–151
gray scale 13, 33, 60
 –, patterns 34

ground calibration 166–167
ground check 59
ground data 103
ground resolution cell 22, 27
ground sampling 105
ground water 13, 37, 54, 87, 93, 116, 145, 165
 –, contamination 33, 41, 42, 89
 –, seepage 111
 –, springs 87

H

heavy metals 129, 140–145, 165, 175
hydrothermal alteration 147, 149–154
hyperspectral scanners 28
hyperspectral sensors 115

I

image classification 108
 –, unsupervised 117
 –, mineral 131–133, 153–157, 170–172
imaging spectrometers 28–30, 156–169
 –, *AVIRIS* 28–30, 115, 116, 123, 131–136, 143, 152–159, 162, 166, 167, 171–178
 –, *CASI* 28, 29, 112, 113
 –, *DAIS* 28
 –, *HYMAP* 28
imaging spectroscopy 164–169
instantaneous field of view (IFOV) 21, 22
interpretative criteria 33–37

K

KVR-3000 10

L

landfill 37, 64–74, 77
 –, abandoned 37, 88, 89, 95
 –, leachate 43, 82
 –, seepage 42, 58, 72, 91, 93
 –, slope instability 49, 51
Landsat Multispectral Scanner (MSS) 166, 168

Subject Index

Landsat Thematic Mapper (TM) 10, 106–114, 123–131, 143, 161, 162, 166, 168
land use 33

M

Middle Ages settlement 54
mineral weathering 140
mineralization 118, 121, 138, 140, 142, 145, 148, 150, 154, 159
minerals
 –, alunite 117, 146, 157, 171, 175
 –, alunogen 117, 146, 156–159
 –, amorphous 171
 –, clays 116, 128–132, 171
 –, goethite 131–145, 171
 –, gypsum 111
 –, hematite 131–145, 171
 –, illite 130–145
 –, iron oxides 116, 127–131, 171
 –, jarosite 117, 131–146, 157, 171
 –, kaolinite 130–145, 171
 –, limonite 127, 132
 –, manganese oxides 133, 143–145
 –, montmorillonite 171
 –, muscovite 172
 –, phyllosilicates 172
 –, pyrite 136–138, 159
 –, smectite 130–145
mine waste 142, 143
 –, characterization 131
 –, inventorying 114, 115
monitoring 106–113
morphology 33
multispectral cameras 12
multispectral image 101
multispectral scanners 10, 22, 23, 99
multitemporal analysis 34
multitemporal investigation 38

N

nuclear waste 96–105

O

Oak Ridge National Laboratory 96

–, waste sites 96
optical-electronic scanners 26, 27
 –, instantaneous field of view (IFOV) 21
optical-mechanical line scanners 21–26
 –, data recording system 22, 23
 –, *Deadalus* series 23, 24, 99, 101
 –, detector unit 22
 –, field of view (FOV) 22
 –, instantaneous field of view (IFOV) 21
 –, line scanning system 21, 22

P

permeability 45, 91
 –, soil 82, 83
PIMA-II infrared spectrometer 115, 117, 135, 144, 151–159
pixel 21, 22
prioritizing 161–163

R

radar 31
Radarsat 10
radioactive waste 96
ratio composite 116
red edge 178, 182
relief 34
reflectance spectra 134, 137–139, 169–175
remote sensing 1, 5
 –, active systems 5
 –, aerial photographs 11
 –, aircraft-based methods 12
 –, nonphotographic sensors 20–30
 –, passive methods 5
 –, radar-based methods 31
 –, satellite-based methods 9–11
remote sensing data 59
resolution
 –, spatial 9, 10, 13, 22, 61
 –, spectral 61
revegetation 108, 113

S

Schoeneiche waste disposal site 73–95
  –, aquifer 73
  –, geologic setting 74
  –, landfill 73, 82
  –, mapping vegetation 91
sediments 172–175
senescence 176, 179
  –, mapping 182
site characterization 1, 33, 36, 73, 75, 111, 143, 161
site prioritization 114, 163
slope erosion 113
smelter 106
soil 6, 75, 82
  –, characteristics 13, 14, 29, 46, 89
  –, contamination 33, 89
  –, moisture 8, 14, 15, 31, 43, 82–85, 87, 88, 103, 105
  –, temperature 8, 77, 79, 82, 103
solar radiation 6
  –, energy spectrum 7
  –, interaction 6, 7
spectral characterization 114
spectral library 156, 176
SPOT 10, 11, 161, 162
Stefan-Boltzmann law 7
stress mapping 183
Sudbury mining district 106–113
  –, Copper Cliff tailings 111–113
sulfur dioxide 106–111
Summitville open-pit mine 164–185
  –, heap-leach techniques 164
surface-water
  –, conditions 53, 165
  –, turbidity 53

T

tailings 40, 106–113, 119, 123, 151, 152, 155–158, 165
temperature 8, 35
  –, blackbody 7
  –, diurnal cycle 35, 41, 77
  –, drainage system 49
  –, earth surface 8
  –, landfill surface 78
  –, sloping ground 78
  –, sun 8
Tetracorder analysis 167, 168, 176, 178, 180, 181, 183
thermal image 41
  –, abandoned landfill 89, 90
  –, drainage systems 48, 49
  –, sloping ground 78
  –, soil moisture 83, 84, 88
  –, tailings 40–42
  –, trenches 100
  –, waste disposal site 78–81
thermal radiance 103, 104
thermal radiation 7
thermal remote sensing 25, 35, 42, 77, 99
  –, buried waste 96
  –, smoldering pyrite 42
  –, surface water 80
thermal scanner 24–26
  –, *AGEMA* thermography system 24, 25, 77
  –, *Daedalus* scanner 99
trench map 98, 102
turbidity 15, 53

V

vegetation 15, 34, 175, 183, 184
  –, anomalies 34
  –, change 109
  –, crops 176–182
  –, damage 106–111
  –, senescence 179–184
  –, stress 91, 93, 176, 182
  –, succession 143
  –, vitality 15, 52, 91
vegetation anomaly 34
verification
  –, in-situ verification 59
  –, remotely sensed data 59
  –, vegetation data 62
  –, virtual verification 59, 60
vitality of trees 15, 91

W

waste disposal sites 1, 9, 14, 19, 33, 38,
    44, 55, 58, 63
    –, Eulenberg/Germany 64
    –, Schoeneiche/Germany 73
    –, search for 58
    –, subsurface 55, 64
waste minerals 126–145
waste trenches 96–105
waste trench map 98

Printing: Mercedes-Druck, Berlin
Binding: Stürtz AG, Würzburg